ISBN 978-0-428-22612-1
PIBN 10276457

1 MONTH OF
FREE
READING

at

www.ForgottenBooks.com

By purchasing this book you are eligible for one month membership to ForgottenBooks.com, giving you unlimited access to our entire collection of over 1,000,000 titles via our web site and mobile apps.

To claim your free month visit:

www.forgottenbooks.com/free276457

English
Français
Deutsche
Italiano
Español
Português

www.forgottenbooks.com

Mythology Photography **Fiction**
Fishing Christianity **Art** Cooking
Essays Buddhism Freemasonry
Medicine **Biology** Music **Ancient
Egypt** Evolution Carpentry Physics
Dance Geology **Mathematics** Fitness
Shakespeare **Folklore** Yoga Marketing
Confidence Immortality Biographies
Poetry **Psychology** Witchcraft
Electronics Chemistry History **Law**
Accounting **Philosophy** Anthropology
Alchemy Drama Quantum Mechanics
Atheism Sexual Health **Ancient History**
Entrepreneurship Languages Sport
Paleontology Needlework Islam
Metaphysics Investment Archaeology
Parenting Statistics Criminology
Motivational

BULLETIN

OF THE

Scientific Laboratories

OF

DENISON UNIVERSITY.

VOLUME IX.

PARTS I AND II.

EDITED BY

W. G. TIGHT, M. S.,

DEPARTMENT OF GEOLOGY AND NATURAL HISTORY.

GRANVILLE, OHIO.

December, 1895—March, 1897.

TABLE OF CONTENTS.

VOL. IX.

PART I.

With 2 Maps.

PART II.

With 11 Plates and 5 Figures.

73/كلل،

BULLETIN

OF THE

Scientific Laboratories

OF

DENISON UNIVERSITY.

VOLUME IX.   PART I.

WITH TWO MAPS.

EDITED BY

W. G. TIGHT, M.S.

Department of Geology and Natural History.

CONTENTS.

GRANVILLE, OHIO, DECEMBER, 1895.

PUBLISHED MONTHLY. COMMENCED JANUARY, 1888

The American Geologist.

ILLUSTRATED.

A Strictly Geological Journal.
Two Volumes of 400 Pages each, per year.

Subscription, ... *$*

Sample Copies 20 cents.

EDITORS AND PROPRIETORS.
Samuel Calvin, Iowa City, Iowa.
Edward W. Claypole, Akron, Ohio.
John Eyerman, Easton, Pa.
Persifor Frazer, Philadelphia, Pa.　　Charles E. Beecher,
M. E. Wadsworth, Houghton, Mich.　　Warren Upham, Somerville, Mass.
John M. Clarke, Albany, N. Y.　　　Edward O. Ulrich, Newport, Ky.
I. C. White, Morgantown, W. Va.
Francis W. Cragin, Colorado Springs, Col.
Newton H. Winchell, Minneapolis, Minn.

PUBLISHED AT MINNEAPOLIS.

BULLETIN

OF THE

Scientific Laboratories

OF

DENISON UNIVERSITY.

VOLUME IX.

PART I

WITH TWO MAPS.

EDITED BY J

W. G. TIGHT, M.S.

Department of Geology and Natural History.

GRANVILLE, OHIO, DECEMBER, 1895.

I.

THE PALAEOZOIC FORMATION.

W. F. Cooper.

No subject open to our investigation is more interesting or economical than that department of inorganic geology embraced under its stratigraphical relations. The conditions attending the origin and deposition of the earth's crust are at the basis of our existence in a certain sense; that is leaving out the ethical element which it is hoped one may consider as the pre-existing state necessary for such physical operations as should best fit man for his abode throughout his entire cycle of life.

Whatever may be the relations sustained by our planet with the rest of the solar system, through the so-called nebular hypothesis; it is suggestive to follow out the comparisons of W. Prinz published in the *Annuaire de l' Observatoire Royal de Bruxelles* for 1891, in which the great continental torsions of the western coasts of America, Europe and Africa, western Siberia through the corresponding coast line of Australia, together with a fourth supposed by him to be indicated by the great chain of islands to which the Marshall group belongs; has been thought to be analagous with similar oblique lines observed on Mars, and less distinctly on Venus and Jupiter.

The similarities afforded by each of the three great continental systems is suggestive as bearing upon the similar primary condition attending their origin and fundamental development. We have counterparts in the respective irregular triangular outlines of North America, Europe, and Asia, in connection with the formation of South America, Africa, and Australia, all with a more or less triangular development, and with the apex of the triangle pointing southward. Similarly as the outlines become less triangular and larger from west to east, we have regularly separating bodies of water also increasing in size, and represented by the Gulf of Mexico, Mediterranean, and the Indian Ocean. The existence of greater depths in the western Atlantic and Pacific, in connection with corresponding altitudes on both of their immediate shore lines, may indicate a great law of stratigraphic equivalence or equilibrium, through which as we already know, great accumulations result in areas of elevation, while vice-versa, lesser deposits might be held as forming the lowest depths. James D. Dana has attributed the zigzag arrangement of continents to torsion with the maximum torsion represented by a belt of volcanoes, and the earth's feature-lines as consequences in part of the pressure or tension attending torsion.

In our own country it can be readily shown that the North American continent had attained the outlines of its present form at the close of the Archaean age, subsequently developing southward into the Appalachian axis; that westward from there the surface deposits prove that it was coexistent with Carboniferous time, while west of a line drawn from the eastern longitude of Dakota the great Mesozoic and Tertiary deposits took place, the great climax of physical activity resulting in the evolution of the Mesozoic age, together with lesser resultant actions at the close of the Tertiary, or during a period very nearly contemporaneous with the Glacial epoch. Coexistent with this, it should also be observed that deposition took place from the north to the south, leaving the southern states adjacent to the Atlantic, Gulf of Mexico, and the Rio Grande of Tertiary and other later formations. We have attempted to show in a very general way the course of development among the sedimentary rocks, and it is thought, that as the general development has been westward, so we may be able to indicate the origin more exactly of portions of the Palaeozoic strata from sources eastward of their main deposition.

Any consideration of the origin of the later fragmentary rocks, involves not only an account of the adjacent land areas, but also the agencies by which they might be removed. The argument of Mr. Bull that the denuding action of tides must have been much greater among the rocks under discussion, cannot apparently find any very strong substantiation in nature, either from an organic or physical basis. Mr. Bull supposes that on account of the greater proximity of the moon at this time, that the action of tides would be greatly increased, causing material to be eroded and deposited in a manner almost inconceivable at the present. There are three objections to this theory: Primarily, that the nebular hypothesis of Laplace involves the fundamental idea that heat is evolved as a result of contraction, not taking into account that the intense heat of the sun would be more apt to cause an expansion instead of contraction along its diameter, according to all the known laws of Physics, while this involves indirectly the relation of our satellite to the earth, and it to the sun. There is moreover, no physical action apparent, with the exception of some cases which will be observed, which would denote any violent physical force. Most of the strata of the Palaoezoic and Mesozoic were deposited during periods when life was very abundant, and its manner of preservation and more particularly of deposition show that the conditions were very quiet, and probably of long duration. In this connection it might also be well to remark that the very nearly equible temperature of the globe which allowed the same animals and plants to flourish on the equator and the Arctic zone at the same time, even as late as the Tertiary period, would also prevent the formation of oceanic currents on a magnitude equal to the present streams, but then as at present, winds acted in promoting such agencies. Another agent which has suggested itself is earthquake waves originating beneath the ocean. We know that the transporting power of water varies as the sixth power of the velocity, consequently if the velocity be increased ten times, the transporting power is increased 1,000,000 times. It has also been ascertained that water moving at the rate of three feet per second will carry angular stones the size of a hen's egg. What would be the result then of a wave 300 miles in diameter, and sixty feet high, moving at the rate of 370 miles per hour in its erosive action upon the adjacent coasts? One can readily conceive that it would be possible to carry boulders two feet in diameter a considerable distance, while the beds of conglomerates which exist in Scotland could be produced by this agency instead of the direct intervention of glacial action as Croll has supposed. We have good reason to believe that earthquake action was as frequent and extensive in the times under consideration as at present, and the great sea-wave just described, which took place during the South American earthquake in 1868, would probable be surpassed by those of previous epochs. Rivers also operated to a large extent, especially during the lower Carboniferous.

Among the elements necessary to the formation of sandstones, and as we shall also consider more particularly of conglomerates, are primarily the elevation of land areas above water as the Archaean rocks of Canada at the beginning of the Palaeozoic age, with other narrow ranges running southwestward parallel with the Atlantic, and still additional areas now represented by the Cordilleras. We have also to take into account that the amount of carbon dioxide in the atmosphere during early geological ages, exerted in connection with water a much more powerful and quickening effect in atmospheric and atmospheric-aqueous action, which must have greatly hastened the denuding action during Silurian, Devonian, and early Carboniferous ages, at the same time changing the chemical arrangement and physical form of the rocks. The lowest orders of plant and animal life also furnished contributions, which taken in connection with the large amount of organic material represented in some limestone formations, as for instance the Hudson group at Cincinnati Ohio, and the carbonaceous elements of many of the black shales constituting the Devonian and the relatively thin, but very important coal seams, clearly indicate the manifold operations of organic existence, as well as theinorganic. In addition to this, we have a counterpart to the formation of coral reefs at the present, duplicated to an unusual extent during the Niagara and lower Devonian, giving rise for example at Louisville, Ky., to a barrier which causes the falls of the Ohio.

An element involving both physical and organic connection is also paramount, as furnishing an index as to the position occupied by the Atlantic ocean. It seems quite apparent that since areas on both sides of its present basin have similarly equivalent, recurring faunas, often quite restricted as in the Cuboids zone, that it was influenced by physical environments which may have also operated in producing sediments for the adjacent coasts, but of this nothing can be said with certainty. Recent surveys have determined the position of three submerged mountain ranges running north and south in its central basin. It is certain that at least portions of the Mediterranean have been eroded to an enormous extent, producing material for the adjacent coasts. Before attempting to trace the origin of some of the sedimentary rocks subsequent to the Cambrian it will be necessary to determine the land areas existing in Ohio. That the Cincinnati geanticine existed at the close of the lower Silurian, forming an island in southwestern Ohio and the adjacent parts of Indiana and Kentucky, is indicated by the absence of upper Silurian and lower Devonian over that area, these formations being deposited on its margins northward. In Tennessee a hiatus is revealed on account of the Devonian black slate resting directly on Lower Silurian beds, clearly indicating the land ares during the upper Silurian and part of the Devonian. This land area had a great influence in building up the subsequent Palaeozoic rocks, as we shall see further on.

The fluctuations and arrangement accompaning the formation of the lower Helderberg strata in eastern New York, show that the upward movement begun there at the close of the Cambrian period, still further progressed after the deposition of the Hudson group, throwing the rocks of that formation above the level of the ocean into anticlinal and synclinal folds east of the Hudson, while decreasing in intensity westward. The Hudson formation may have furnished sediment from which the Oneida comglomerate was in part derived, but it is apparent that the later beds of the Niagara period had a connection with the vastly thicker formation in Canada through a channel possibly leading northeastward from western New York. With only a thickness of 300 to 400 feet in Ontario, increasing to 1,300 feet in Nova Scotia, it seems possible that the sediment was derived from regions north or northeast of New England, while the intimate relation of its fauna to that of England point to a very close biological relationship between the two areas, which oftentimes results from an uniformity of physical environments. The thinning out westward of the lower Helderberg group in New York, together with its comparative thinness in Tennessee, demonstrates it to be an essentially eastern formation. Unlike the Niagara, however, it thins out very rapidly to the westward until in Cayuga county it has almost entirely disappeared. It is obvious, however, to the most casual observer, that the Helderberg escarpment in Albany county must have had a much greater extension northward than at present, and after H. Fletcher's determination of the thickness in Nova Scotia (1,000 feet), we can probably admit the truth of Logan's determination that it was connected with the Canadian basin through the Champlain valley; bounded on the east and west by the folds of the Cambrian and the Adirondack range.

Continuing upward in the geological formations we find the Hamilton group with a thickness of 1,200 feet in the Catskill region, but rapidly thinning to the westward, until in western New York it is scarcely 200 feet thick, while at the falls of the Ohio the beds include 20 feet of impure limestone. In eastern Pennsylvania the greatest maximum thickness is 5000 feet, in the Gaspe region 6000 feet. The associations of the specimens I have seen from Perry (?) Maine would indicate an estuary connection with the Gaspe fauna, outside of the main line of deposition. The manner of preservation is very similar to specimens from western New York. It is impossible not to believe that the Hamilton strata did not extend farther eastward, northward, and northwestward than at present, but all the strata have suffered denudation to an enormous extent, and we would not know of any northwestern connection, but for the Mackenzie river deposits. It is possible that portions of the Hamilton formation were derived from uplifted beds of the Hudson group and other formations east of the Hudson valley, and the Adirondacks may have contributed its share. It is obvious that the sediments could not have been derived from either the west or the southwest, and since according to Dana's determination the Champlain outlet was closed it is apparent from the relative thickness in the Catskill region, Pennsylvania and Nova Scotia, that the parent rocks may have been what is at present the bed of the western north Atlantic ocean, but it is entirely hypothetical. The lithological aspect of the formation is very suggestive as to its origin, especially when taken in connection with the organic remains. In eastern New York the strata are silicious with interspersed beds of shale, and containing land plants very similar to those described by Dawson from St. John. Lepidodendra as drift material, together with Psaronius actually growing and covered by the deposits. Farther westward the strata become thinner and more argillaceous, indicating quiet marine conditions, and greater distance from the source of the sediment.   Certainly, however, there was an open connection with the eastern Canadian basin, by means of which an active inorganic and to a lesser extent faunal relation was sustained.   It is also apparent that the strata in the Gaspe region were much nearer the original source of the sediment.   I have rarely noticed very thin conglomerate beds in the Schoharie valley suggesting shore-line deposits—the precursors of greater strata which were deposited in the Chemung and Carboniferous strata.

Scarcely any subject in Palaeozoic stratigraphy with the exception of the Taconic question has caused more discussion than the relation of the Chemung, Catskill, and Waverly formations. Alexander Winchell would have had us believe that the Waverly and Catskill were in the same basin of deposition and coexistent. Another author contends, and very probably, that the Chemung and Catskill are equivalent formations with only a lithological difference due to different physical environments, while the Chenung and Portage are related but distinct formations.   Prof. J. M. Clarke from palaeontological evidence links the Portage with the underlying beds.   Prof. J. J. Stevensen on the other hand uses Chemung for a generic term with the divisions Portage, Chemung and Catskill, and finds that the Catskill period presents a closely circumscribed area during the deposition of the last beds of the Chemung but was greatly enlarged to the southward during the formation of its upper beds. We incline to this opinion, at the same time correllating the Chemung of Brown county N. Y., with part of Mather's Catskill group of the Catskill mountains as Mr. N. H. Darton has done, leaving the Catskill group of Stevenson as a formation which had its greatest and typical development south of the Pennsylvania line, and outside of the typical locality which includes strata not understood when Mather made the survey of his district. It may be that the red Bedford shale lying at the base of the Waverly formation in Ohio represents a connection with the Catskill of the east about the latitude of Pittsburg, but it contains recurrent Hamilton species which lingered in the west long after the Hamilton formation was succeeded by later deposits in New York. The Bedford sea could probably be represented as an estuary in Ohio, which was bounded on the west by the fold of Cincinnati rocks and those of later age. The gradual uplifting of northern Ohio which had then begun, continued in operation until the following lower Carboniferous horizons were deposited on a shoreline which steadily progressed southward.

The Chemung group is 350 feet thick in southwesthern Virginia on the Tennessee line, but rapidly thickens northward, being 3800 feet thick on the boundary line between Virginia and Pennsylvania, 4700 feet in Huntington county Pennsylvania, and four thousand feet near the New York line on the Delaware river, while in the Catskill mountains it is 3000 feet thick. In southwestern New York the Chemung is 1200 feet thick. The Chemung strata thin out to the westward and south-westward in Pennsylvania, but northward along the western boundary line it reaches a thickness of 1400 feet in Crawford county near Lake Erie. The Erie shales which represent the western extension of the Chemung in Ohio rapidly thin out as we approach Columbus, almost, if not quite disappearing as we approach that city. Prof. E. Orton states its thickness at 300 feet, but it is very variable. When the stratigraphy is subject to so much variation in thickness, lithological appearance, and distribution, we must be prepared to be somewhat at variance concerning the origin of its individual beds of conglomerates, as well as the remaining strata. Prof. I. C. White has correllated the Panama with the Alegrippus conglomerate,

and likewise the Salamanca layer with the Lackawaxan pudding-
stone, thus making the two layers continuous over southwestern New
York, across eastern Pennsylvania on a line rudely parallel with the
Blue ridge into southwestern Virginia. In the Catskill mountains a
layer of conglomerate is present which may be equivalent to one of
these layers. It seems quite natural that the conditions attending the
formation of the Hamilton group again operated to a lesser extent
during the upbuilding of the Neodevonian. It cannot be denied that
the lower Chemung was restricted in its basin to the northward, while
in its limited extension east of the Hudson the highlands of New
England may have furnished material for its upbuilding. In
Ohio rather abruptly limited on the west by the Cincinnati
uplift, the strata were deposited in a sea whose main axis ran
north and south, and which received its sediment from currents directed northward. It may possibly be that the hiatus existing in Virginia between the Archaean and Tertiary went to supply part of this
material, while it is natural to conceive that strata now only represented by the West Indies may have not only sent its contribution to
this formation but still others in the geological scale. But of that
nothing can be said with certainty. The effect of tidal currents is
essential in producing such beds of conglomerates as have been laid
down during this age, and in the flat and round pebbles of the Panama
and Salamanca conglomerates, we have illustrated the active erosive
in shallow waters, with the result of somewhat different physical
environments.

Although somewhat intimately related in the eastern extension of
its basin with its underlying rocks the deposition of the Waverly
shales in Ohio witnessed an important change in the physical geography of the lower Carboniferous formation. On the northwest there
was a close connection with the Marshall group in Michigan, and
even after the Berea and Cuyahoga shales had been uplifted in north-
eastern Ohio the channel remained open at least until the middle
Waverly freestones had begun to be deposited farther south, probably
receiving accessions for growth from the Cincinnati geanticline, but
the physical conditions for faunal existence farther south toward
the Ohio river were not favorable at the close of the lower first division of the Waverly. East and northeast from central Ohio, the conditions attending the deposition of the middle and upper Waverly
were apparently more favorable under the coal bearing formations

south of lake Erie, and it is only in that way that we can account for the occurrence of conglomeritic beds and other higher horizons, which are not represented in northern Ohio, but which nevertheless reappear in Pennsylvania. To Prof. C. L. Herrick we are greatly indebted for the determination of the different zones of the Waverly, and more particularly of their faunal characteristics. The student desiring to obtain some conception of the evolution of biological forms from the Devonian to the Carboniferous, together with an exact idea of the stratigraphic relations in this formation, is referred to the Bulletins of the Scientific Laboratories of Denison University volumes III-V, and volume VII of the Ohio State Geol. Survey. While the increasing thickness of the upper Waverly formation toward southern Ohio points in that direction as from which sediments were borne, we are apparently confronted with the fact that the rivers which deposited the conglomerates came from the northeast. They differ from the beds forming the Chemung conglomerates in the comparative restriction of their areas and manner of deposition, but are so closely allied lithologically as to point towards a common source of origin. In the Waverly all that we know definitely of the upper conglomerate which may represent a repetition of the lower member is that it was deposited in the deepening sea near Portsmouth, Ohio, and along a north and south line east of Newark, through Mount Vernon, near Independence which is about twelve miles southeast of Mansfield, where it was at one time confused with the Carboniferous conglomerate 150 feet higher, and having on both its east and west sides broad and gradually decreasing deposits. A section at a right angle to this line of deposits at Lyon's falls near Independence shows a layer forty feet thick thinning out very rapidly east and west. At the "Back bone" one mile east it is only two feet thick, while it entirely thins out westward. It disappears under the coal measures northeast of Wooster, and we are left to speculate as to its further course, and origin. It may be that some of the sub Olean conglomerates in northwestern Pennsylvania belong to the same horizons, but this remains to be verified.

The beds of sandstone overlying conglomerate II near Black Hand occasionally contain pieces of chert, which is characteristic of the St. Louis formation at New Providence Indiana. It is very suggestive as showing the direction taken by some of the currents which deposited the upper Waverly, besides furnishing an index as to the age of those sandstone beds underlying the Chester or Maxville limestone.

The great salient features concerned in the growth of the American continent are represented by the Archean nucleus with its essential importance in building up strata and protecting the life of the seas washing its shores ; the Cincinnati geanticline further modifying and hastening the processes long since inaugurated, and forming a basin which exerted a profound influence not only physically but biologically; and finally the Appalachian revolution which terminated to a great extent courses which had long been in operation for the preparation of the world for the coming of man, but which, nevertheless, made tangible the physical environments necessary for the higher existence.

II.

LICHENS OF LICKING COUNTY, OHIO.

J. Orrin R. Fisher, M.S.

The following list is intended to include such lichens as have been found in this county, and worked up by the author during the school year of 1893-94 in the Denison University laboratories. To this list have been added a few specimens from the neighboring county of Muskingum; the localities given being followed by " L." for the former, and by " M." for the latter county.

The names and authorities are given in conformity with Tuckerman's "Synopsis of the North American Lichens," and most of the identifications have been confirmed by other investigators.

In this list are at least four specimens not before reported for the State, viz:

Cladonia symphycarpa Fr.;

Lecanora cenisia Ach.;

Pannaria nigra (Huds.) Nyl.;

Peltigera canina (L.) Hoffm., var. spongiosa Tuckerm.

As to collecting it is difficult to say where are the best localities, for all are good; but a trip to Black Hand for crustaceous and rock lichens, to Buckeye Lake or Munson's Hill for fruticulose lichens, and to the region around Newark or to Pleasant Valley for foliaceous ones, will reward the collector with many fine specimens.

The study of lichens is an interesting one, and offers to the diligent and careful student a rare field for investigation from the fact that there are many points yet to be determined in their structure, and very many specimens yet to be classified and named.

The following lines quoted by Henry Willey in his " Introduction to the Study of Lichens," represents the attractiveness of the study:

"If I could put my woods in song,
And tell what's there enjoyed,
All men would to my garden throng,
And leave the cities void.

In my plot no tulips blow ;
　Snow-loving pines and oaks instead;
And rank the savage maples grow,
　From Spring's first flush to Autumn red.
My Garden is a forest ledge,
　Which older forests bound."

*　　*　　*　　*　　*　　*　　*

"Wings of what wind the Lichen bore,
Wafting the puny seeds of power,
Which, lodged in rock, the rock abrade ? "

LICHENES.

TRIBE I. PARMELIACEI.

RAMALINA, Ach., DeNot.

R. CALICARIS, (L.) Fr., var. FASTIGIATA, Fr. Granville, L., and
Adamsville, M.   Occurs on trees, &c., in moist localities.

USNEA, (Dill.) Ach.

U. BARBATA (L.) Fr., var. FLORIDA Fr. Buckeye Lake, L.,
Adamsville, M.   On trees and dead wood, abundant.

THELOSCHISTES, Norm. Emend.

T. CONCOLOR (Dicks) Tuckerm. Granville, L. Common on
trees and rocks.

PARMELIA, Ach., DeNot.

P. BORRERI Turn., var. RUDECTA Tuckerm.   Granville, Buckeye
Lake, L.

P. CETATA Ach.   N. E. of Granville, L.

PHYSCIA (DC., Fr.) Th. Fr.

P. SPECIOSA (Wulf., Ach.) Nyl.   Fair grounds, Newark, L.   On
trees.

P. HYPOLEUCA (Muhl.) Tuckerm.   Spring Valley, Granville, L.

P. LEUCOMELA (L) Michx.   Buckeye Lake, L.

P. AQUILA (Ach.) Nyl., var. DETONSA Tuckerm. Toboso, or
Black Hand, L.

P. STELLARIS (L.) Tuckerm.   Granville, L.

P. TRIBACIA (Ach.) Tuck. herb.   On Newark Fair Ground, and
near Haven's Quarry, S. of Newark, L.

P. HISPIDA (Schreb., Fr.) Tuck. herb. On branches in N. side of Cranberry Marsh, Buckeye Lake, L.

STICTA (Schreb.) Fr.

S. AMPLISSIMA (Scop.) Mass. Granville, L, Abundant.

S. PULMONARIA (L.) Ach. Common in Licking county, but none found with apothecia. A specimen collected in a strip of dry woods one mile N. of Sonora, Muskingum county, exhibited apothecia containing the typical cymbiform spores.

PELTIGERA (Willd, Hoffm.) Fée.

P. HORIZONTALIS (L.) Hoffm. Munson's Hill near Granville, L., on the earth; near Adamsville, M., on mossy rock.

P. RUFESCENS (Neck.) Hoffm. Toboso, L. Under a rock ledge.

P. CANINA (L.) Hoffm., var., SPONGIOSA. Tuckerm. Granville, L.; Pleasant Valley, M. On the earth and dead-wood.

PANNARIA, Delis.

P. NIGRA (Huds.) Nyl. Toboso, L.; Pleasant Valley, M. On large sandstone rock.

PLACODIUM (DC.) Naeg. & Hepp.

P. FERRUGINEUM (Huds.) Hepp. Very abundant on small stones in the vicinity of Granville, L.

P. FERRUGINEUM (Huds.) Hepp., var. DISCOLOR Willey in Litt. Granville, L.

LECANORA Ach., Tuckerm.

L. CENISIA Ach. Granville, L.

L. SUBFUSCA (L.) Ach. Granville, L. on wood and rocks.

L. VARIA (Ehrh.) Nyl. Granville, Buckeye Lake, L.; Pleasant Valley, M. On trees, well distributed in both counties.

RINODINA Mass., Stizenb., Tuckerm.

R. SOPHODES (Ach.) Nyl , var. EXIGUA Fr. Pleasant Valley, M., on bark of trees.

TRIBE II. LECIDEACEI.

CLADONIA Hoffm.

C. SYMPHYCARPA Fr. Granville, L., Zanesville, M., on the earth abundant.

C. MITRULA Tuckerm. Granville, L.

C. PYXIDATA (L.) Fr.   Granville, L.

C. FIMBRIATA (L.) Fr.   Granville, L.

C. GRACILIS (L.) Nyl., var. VERTICILLATA Fr.   Munson's Hill near Granville, L.

C. SQUAMOSA Hoffm.   Granville, L.

C. DELICATA (Ehrh.) Fl.   Granville, L., Zanesville, M.

C. FURCATA (Huds.) Hepp., var. CRISPATA Fl.   Granville, L.

C. RANGIFERINA (L.) Hoffm.   Toboso, L., on the earth upon Black Hand Rock.

C. RANGIFERINA (L.) Hoffm., var. ALPESTRIS L.   Locality same as preceding.

C.CRISTATELLA Tuckerm.   Granville, L.

C. UNCIALIS (L.) Fr.   Toboso, L. on Black Hand rock.

C. CÆSPITICIA (Pers.), Fl.   Nashport, M.

C. RAVENELII Tuck.   "Alligator Hill" near Granville, L. on board fence.

BEOMYCES Pers., DC.

B. ROSEUS Pers.   Toboso, L. on the earth upon Black Hand rock.   Rare, having been found in no other locality.

LECIDEA (Ach.) Fr. Tuckerm.

L. ALBOCÆRULESCENS (Wulf.) Schaer.   Toboso, L.

L. PLATYCARPA Ach.   Toboso, L.

BUELLIA DeNot., Tuckerm.

B. PARASEMA (Ach.) Th. F.   Pleasant Valley, M. on trees.

B. PETRÆA (Flot., Koerb.) Tuckerm.   Granville, L.

TRIBE III.   GRAPHIDACEI.

GRAPHIS Ach., Nyl.

G. SCRIPTA (L.) Ach.   Granville, L. abundant on dead wood.

TRIBE V.   VERRUCARIACEI.

VERRUCARIA (Pers.) Tuck.

V. RUPESTRIS Schrad.   Fultonham, M.

PYRENULA Ach.

P. NITIDA Ach.   Granville, L.

III.

A CONTRIBUTION TO THE KNOWLEDGE OF THE PRE-GLACIAL DRAINAGE OF OHIO.

PART II.

Pre-Glacial and Recent Drainage Channels in Ross County, Ohio.

By Gerard Fowke.

Ross county presents an interesting field for the student of glacial geology.

The southern limit of the ice-sheet is marked by a well-defined terminal moraine which follows almost exactly the diagonal of the county, as it enters at the northeast corner near Adelphi and passes out about two miles beyond Bainbridge at the junction of Ross, Pike, and Highland counties. There are few points along this line where the drift is not a prominent feature of the landscape; in many places it has a thickness of more than 100 feet exposed and occasionally attains an elevation of about 150 feet above the streams which flow across it or along its margin. Some very large "kettle-holes" exist on this border; while numerous sections along the nearly vertical banks of streams or in excavations for ballast or pike material afford excellent opportunities for observing the complicated structure produced both by the ice itself and by currents from its melting. These features, however, except perhaps as to the thickness of the deposits, are common in all glaciated regions and may be as well studied elsewhere; but there are few, if any, places where in an equal area may be found so great an alteration in water courses as has taken place in the southwestern quarter of this county since it was first invaded by the glacier.

By reference to the map (Plate I), it will be seen that, at present, Paint creek forms the western boundary of the county from Greenfield to the mouth of Rocky Fork, near the point marked *H*. Thence it flows nearly east for about three miles, after which its general direction is northeast to the point *E*. Here it bends abruptly to the southeast, then toward the northeast to the point *C*, from which its general course is east to the Scioto river. Somewhat more than two miles above Rocky Fork, a considerable tributary, Rattlesnake Creek, or Rattlesnake Fork, enters ; and the enlarged stream pursues a tortuous course of several miles to *H*. At the point *G* is a ledge of limestone, forming the Falls of Paint Creek. The largest tributary within Ross county is North Fork whose head-springs are on or beyond the Fayette county line; it flows southeast past Frankfort and enters the main stream at *C.* There are many smaller streams, and scores of ravines, some of them several miles in length; but only those are represented which contain water all the year. The lowest level at which the bed-rock is visible, whether in the bed of streams or on its line of contact with the drift, is shown by heavy dotted lines; no account is taken of the superficial deposit on the table lands, which in most places is quite thin and frequently is altogether lacking. In the low-lands the drift extends to a greater depth than has ever been reached by well diggers. The crossed lines denote drift-filled valleys in which there is now no running water in greater amount than may come from a small spring.

A tour of discovery by a person unfamiliar with the country, starting at Greenfield to follow the course of Paint Creek, would, to judge from the experience of the writer, result somewhat in this way. except that a large part of the territory here figured would have to be closely and accurately examined many times before the facts were understood :

From Greenfield southward the investigator will find limestone cliffs bordering the stream, separating here and there with little level valleys between them, the water skirting the rock along first one side and then the other. Tributary ravines, dry most of the year, show similar gorges or canons. When he reaches Rattlesnake he finds the valley swing east and widen considerably, with heavier deposits of drift; but suddenly it turns southward again through a valley more contracted, with rocky walls. Following these, he presently turns northward, and finds drift deposits on both sides of him. These, however, soon disappear, and he follows a long loop through bed-rock, coming after awhile to a place where the stream again flows through drift almost to the mouth of Rocky Fork. From here, for several hundred yards (at *H*) the bottom of the creek is solid rock, with thousands of "pot-holes" and long narrow grooves cut by the stones and sand whirled along by the rushing water which in the last 200 yards of this rocky bed has a fall of 19 feet. Suddenly the turbulent stream comes to rest in a quiet pool which has a depth near its upper edge of about 80 feet. The right bank continues its course as an unbroken bluff; but the left bank abruptly terminates with a sharp anvil-like point projecting into the deep water. On crossing over, it is seen that the upper edge of the pool, on the northern side, is some distance above this point, with a muddy shore in which no limestone is apparent. This shore gradually curves around toward the east until it forms a bank to the creek parallel to that on the south side. Climbing the gravel hills to a point north of *H*, the traveler sees to the westward conical and roof-like hills, whose smooth-flowing outlines show them to be of drift material, stretching away to the bend just below Rattlesnake which he had left some hours before; and he further sees that they appear to cross at the points where he had lost the limestone on his way down. Thorough examination, involving many miles of tramping to and fro, convinces him that Rattlesnake is flowing in a pre-glacial valley which was filled with drift from the junction of Paint creek to this deep pool at Rapids Forge *H*; and that after seeking outlets in various directions as shown by abandoned channels and minor terraces it finally escaped along its present crooked way, regaining its former bed by cutting out the limestone which had made its southern boundary, washing down-stream the gravel that it found filling the present pool and making with it a dam which retains the water. He finds also, that the beds of both Rocky Fork and that portion of Paint creek above the mouth of Rattlesnake have been eroded in post-glacial times.

Somewhat more than a mile below the pool at Rapids Forge, the rugged hills on the south cease and in their stead appear conical knolls which cause the observer to rub his eyes and wonder if he has been suddenly transported to the region of Omaha; for at no nearer point will he find such remarkable resemblance to the Missouri river bluffs. Next, he sees a valley opening from the south, and then reappear the hills capped with Waverly sandstone such as he had seen above; but they are farther away from him. Following the road along the creek bank he soon approaches their foot; and now the hills on the north side have receded, while the creek, making a salient angle, seems bent on following them. Leaving the road, which continues in nearly a straight course and following the creek to the Falls *G*, he finds a cataract 8 feet high pouring over the last exposure of limestone in the valley, the bed rock from here to the Scioto being shale. In the bottom land, just east of the falls, is a very large gravel deposit, part of the old moraine, with lower ground between it and the hills to the southward. It is apparent that at a comparatively recent date the creek has flowed through, or south of, the site of Bainbridge. More time is required for one fully to realize that he has followed thus far what was once only a tributary to a far larger stream; that Rattlesnake formerly had its mouth just above the Falls; and that only now has he reached the true valley of Paint creek. But the sudden widening from a few hundred feet to nearly a mile; the break in the hills to the southward, filled with gravel-knolls over 150 feet high; the persistence of this gravel up to Beech Flats with rock-capped hills on either side; the width of the valley, almost as great at the Falls as at any point below;—all are proof that the headwaters of Paint must be sought to the southwest, possibly as far away as Brown or even Clermont county, for all the streams rising in the area which may formerly have been drained by this lost part of Paint creek flow southward or westward through gorges or narrow troughs in their whole course, none having the broad valley so characteristic of this. Mr. H. W. Overman, of Waverly, pointed out years ago that the drainage of Ohio Brush creek was reversed, its natural course being to the north instead of to the south. The same will probably be found true of other streams still more to the west.

Leaving this for future determination, our student goes on down the broadened valley, admiring the wonderful fertility of the soil, the fine farms, the picturesque beauty of the sloping or, sometimes, precipitous hills that border it. Perhaps he turns aside at *F* to examine the vertical exposure of nearly 300 feet of shale at Copperas mountain into whose base the creek has cut its way; he notices a dark line near the top which marks the separation between the Devonian and the Subcarboniferous. Similar, but much smaller, sections may be found at other places where the creek cuts against one hill or the other as it swings back and forth across the intervening space. Not far below *F* is a fine vertical exposure of gravel capped with clay and sand, in all about 60 feet; the bank is rapidly caving and is now several yards within the original line of the pike which has been twice moved back. Finally the creek, skirting along the southern range of hills, is lost to sight for about three miles and is next seen at Slate

Mills. But instead of flowing on gravel as heretofore, it is on a bottom, and between banks, of bedded shale, with gravel on the farther side above the shale. While the traveler is pondering over this, he suddenly observes that it is flowing to the right, as it did when he crossed it before, two miles below Bainbridge; and he knows he has not crossed it twice. He looks on, in the direction toward which he is traveling, and sees the same range of hills on either side bordering a drift-filled valley such as he has been coming along for several miles; but he is now looking *up* the stream instead of *down* it. More puzzled than ever he leaves the pike and follows the stream which soon curves around to the westward. He thinks this is as it should be until he unexpectedly finds himself on a railway which he does not remember to have seen; true there was a railway near Bainbridge, but he knows he is not back there; besides it is not going in the same direction. Presently he sees another railway; both of them, with the creek, disappear in a narrow gorge which he certainly has not seen before. Next, he notices that the stream is not more than one-fourth as large as it should be. Wondering if he is bewitched, he climbs a hill, looks to the westward, recognizes numerous places he has passed; looks eastward and sees the continuation of the valley, but without a sign of water in it. He tries to trace the stream he has just left; it passes the bridge where he first saw it, wanders through a narrow valley, runs up to a high hill, and apparently stops there. He then walks south across fields, thinking thus to reach the larger stream, and finds himself at the bridge again. He inquires at the mill near by as to the location of Paint creek, and is told with a vague general flourish of the hand in the direction of the setting sun, that it is "up that way." Retracing his steps for weary miles, he finds his lost stream half-way back to Bourneville. Determined not to lose it again, he notes the trend of the current, starts in the same direction closely watching the hills to the south, and is satisfied there is no place it can pass through. He can not *see* the stream, but he knows it runs along the foot of the hill under those huge elms and sycamores. Soon he finds himself at the foot of the hill near *D*; but there is no creek visible—the gravel is piled up against the slope. Uncertain whether to swear or to pray, he walks on and reaches the mill, whose owner eyes him suspiciously. Making further inquiries, he learns that the bridge is over North Fork, which flows into Paint creek about two miles from where he stands. Taking the new direction to the

southward, he finds Paint creek again at *C*, and follows it thence
through a broad valley to the Scioto bottoms. Coming back to *C*
and ascending Paint creek he observes that the hills on either side
contract in a V shape toward the mouth of Ralston's run, which runs
through a level bottom about 500 feet wide with steep hills on either
side. From here up to a ravine putting in from the west, there is a
strip of bottom land on one side of Paint creek, nowhere wider than
400 feet; and from this ravine up to the point *E* the hills ascend
from the edge of the water which flows on solid rock all the way.
At *E* he finds the creek in its proper channel at the foot of the hill,
under the elms and sycamores, just as he had thought when looking
from the pike.

The order of events that gave rise to these conditions is apparently about as follows :

It is plain that the glacier reached, as a mass, to the old valley
of Paint creek and that it did not ascend the hills on the southern
slope or even reach to them anywhere below the point *F*, except
at the points *B* and *D*. About two miles below Bainbridge, the
drift is piled half-way to the tops of the hills to the south, and the
valley along here must, for a time, have been entirely closed by ice.
There is no doubt that it thus followed the valley nearly or quite to
its head, leaving the deposits above Bainbridge, probably forming
the Beech Flats, and filling up all the valleys when it passed out
upon the limestone table-land beyond the rugged hills of shale and
sandstone; thus deflecting toward the Ohio all the waters which in
this region had flowed into Paint creek. But, as above stated, this
is still to be worked out. At the point *B* where the creek formerly
discharged into the Scioto the drift is fully 100 feet higher than the
highest river terrace; the distance between the hills, measured on
the drift, is a little less than one mile. The flat-topped hills at this
place are about 100 feet higher than the drift. This denotes a sufficient thickness of ice to dam Paint creek and form in its basin a
lake which, fed by the natural drainage and the floods from the
melting ice would rapidly rise until it found a new outlet.

The Scioto having a deep pre-glacial channel, it is very probable
that a lobe closed up the mouth of the creek at *B* some time before
the main body of ice surmounted the hill and filled its bed. Between
Slate Mills and the point *C* there was evidently a low depression
formed by two ravines, one opening north into Paint valley, the

other south into Ralston's run (which then extended to the Scioto river, the creek having usurped its ancient channel) with a low divide between them. At its narrowest point this is now about 1000 feet wide. It is evident at a glance that Paint creek should have turned this way on abandoning its old channel; for this pass, as shown by its width, was much lower than any other that existed anywhere along the southern border. In fact, the stream must have gone through this way for a long time, and with a great volume of water, for it is impossible that so wide a valley could ever have been formed by natural erosion in so short a distance. Thus the whole drainage of Paint creek, reinforced by that from the glacier, would escape through this depression into Ralston's run at a level sufficiently above the Scioto to create a swift current, cutting both the depression and the run wider and perhaps deeper. When the glacier reached its ultimate extension, as a body, within the limits of Ross county, a spur reached entirely across the valley at the point D where the drift is piled to a height of about 120 feet above the present level of the creek. This was from a solid extension and not from a floating berg, because it is pushed up to this level over the solid rock. It would, consequently, shut off the former outlet and form from D southwestward a lake which rose until it began to flow over a col or saddle-back at the point E into a ravine, tributary to Ralston's run, which had worked its way back until it had to some extent lowered the crest in this range of hills; there was no corresponding ravine on the northern side. The cap rock is Waverly sandstone, full of joints; the underlying shale is so loose it can easily be dug out with a pick. When into such material a lake abundantly fed plunges from a height considerably greater than Niagara, the incoherent rock would disappear almost like wax before fire. If the present Ohio was closed at this time, the Scioto was a lake; if the former was open, the latter was a surcharged torrent.* In either case, it was backed up against these hills, forming a body of dead water in which all the rock eroded from the new gorge, along with such material as could be carried by ice, found a resting place, and settled on the drift that had been carried into the same backwater when the larger creek came down from Slate

---

*Since the above was put in type investigations in another direction have shown that the flood-height of the Scioto, immediately below Chillicothe, was at least 200 feet above its present bottom.

Mills.   This we know, because an area of fully a square mile about C has a solid deposit of drift-material rising more than 100 feet above the creek, composed largely of sandstone blocks whose angles are scarcely worn, and masses of shale sometimes containing two or three cubic feet, which disintegrate after a few weeks of exposure to the weather. These could have come only from the gorge between E and the present mouth of Ralston's run.   They are intermingled with sand and northern rock, promiscuously for the most part, but occasionally with a rude stratification as if the floods had been somewhat intermittent. The great apparent width of the valley below this point is due mainly to the filling-in by drift; but, it may be in part, also, to the earlier discharge of Paint creek having enlarged it to some degree, as mentioned above.   The fall of ravines and minor streams outside the glaciated area is rapid; and in those filled with drift the depth to bed rock may be roughly estimated by carrying downward the line of slope of the hills on either side to their point of intersection.   So of the larger streams.   This statement, of course, does not hold good near the junction of two streams; when they have cut down to the level of those into which they flow, smaller streams can not further deepen their beds, but will swing from side to side thus making narrow bottoms.   This is why there is always a widening of the valley where two branches unite; and in such cases the rule just given will not apply.

* * *

From the northwestern point of the hill southeast of Frankfort, one looks to the horizon northward over a practically level drift covered country.   The hill on which he stands bears, on one hand north of east to the Scioto bottoms: on the other, it reaches a short distance southward, bends toward the west, and finally sweeps away northwest until it is lost to sight.   This hill-land and the portion of the plain adjacent to it are drained by North Fork; the part north and east of Frankfort is drained by Deer creek.   Both are of post-glacial date.   The latter, being entirely superficial as regards the drift, need not be considered: the former has a history.

Prior to the advent of the ice, that part of the present valley of North Fork between Frankfort and Paint creek was a depression with an outlet in each direction, the dividing point between the two ravines being near where Union and Twin townships corner.   At this point the shale hills are now less than 100 yards apart; just below (south) they separate considerably, apparently planed out by the ice, and there are drift deposits more than 100 feet high through which the creek now cuts its way. These may have been formed by the advancing ice pushing through the valley a little ahead of that on either side which had to ascend the hills; or they may have been left on its retreat: or it is possible, though scarcely probable, they mark a re-entrant angle of the moraine. It is true there is a wide gap in the heavy deposits in the main stream below here, but it is more reasonable to suppose that they have been removed by erosion than to believe the ice-sheet would stop moving in a place so favorable to its progress. At any rate, a lake of considerable depth was at one time held back above them; for at the point A. on the hill-side, 75 feet above the railway, is a finely stratified deposit of sand. This, however, may have settled in the water which rose in front of the ravine (which then extended much farther to the north,) until it broke over the divide. When this happened, such water as went out this ravine became a part of Paint lake until the extension of the ice confined the latter above the point D. But between B and D there would still result from the melting ice a great quantity of water whose most natural, and indeed only means of escape until new channels were opened miles to the northward, would be toward C along the bed from which Paint creek had been so summarily shut off. This continued until the present course had been cut to a depth lower than the surface toward the east or west; and North Fork, being thus debarred from following the old valley in either direction, has ever since flowed directly across it, high above the original bed, as though carried on an aqueduct.

NOTE.—Too late to add to above paper, I discovered a glacial outlet in the eastern part of Ross county. A number of ravines from the hills back of Mount Logan and Rocky Knob, united and flowing past Mooresville or Halltown, discharged into the Scioto about four miles below Chillicothe. A smaller ravine skirted the northern slope of Rattlesnake Knob, and entered the river not far below the other.

With the greatest extent of the glacier, a lake was formed between its front and the hills a short distance west of Adelphi, over what is now known as Maple Swamp. This finally broke over into the first ravine mentioned, making a narrow gorge in the hills; at the lower end of this there are drift-hills whose summits are at least 150 feet above Walnut Creek which now flows between them and the hills to the east.   These deposits, extending for miles, and uniting with those made by the huge eddy formed by Mount Logan (which forced the glacial currents in the Scioto to the western hills), completely shut off these two pre-glacial ravines, and forced the water coming through the Maple Swamp gorge to skirt the hills, overflow the col back of Rattlesnake Knob, and sever that peak entirely from the range of which it formed a part.   Through the narrow gorge thus made, Walnut Creek finds its way, and reaches the river miles below.

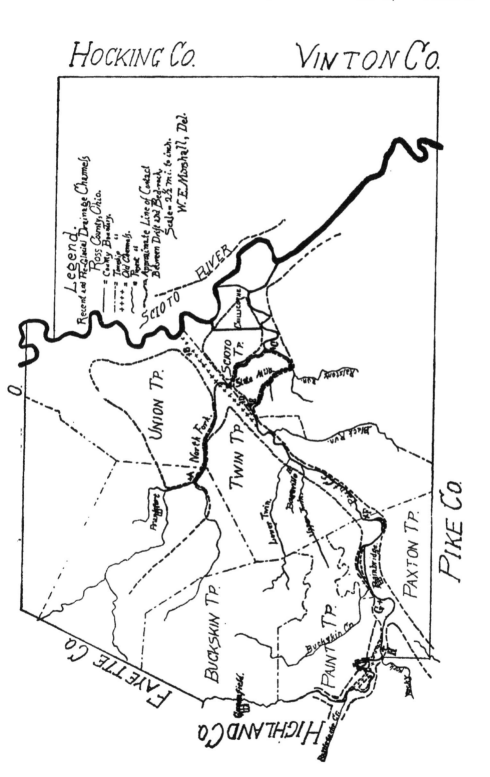

PART III.

A Preglacial Tributary to Paint Creek and its Relation to the Beech Flats of Pike County, Ohio.

By W. G. Tight.

Reference is made in the preceding article, in this volume, on "The Pre-glacial and Recent Drainage Channels of Ross county, Ohio," on page 17, to the extension of the preglacial valley of the upper part of Paint creek to the southwest, at a point a little above Bainbridge. It was with a view to ascertain the course of this tributary to Paint creek that these studies were undertaken.

The results of glacial action along the margin of the ice sheet are so varied and at times so unexpected that almost every acre presents some new and interesting features. This region, lying as it does just on the boundary line of greatest glacial extention, is no exception. While it presents some characters common to some other localities studied, yet there are many new features which add especial interest to this field. A very casual observation revealed the fact that this preglacial channel extends to the Beech Flats of Pike county and is in some way connected with their origin. A view from one of the high Waverly hills at the junction of this valley with Paint creek would easily lead one to the conclu sions stated by the author of the preceding article on page 18.

So striking are the topographical features of this region that we find them mentioned in the earliest writings on geology of this portion of the state. Dr. Edward Orton in his "Report on the Geology of Pike County" in the "Geological Survey of Ohio," Vol. II, 1874, makes the following statement with reference to this locality: "In the extreme northwestern . . . corner of the county, near Cynthiana . . ., there is a conspicuous example of surface erosion that does not belong to either of the systems thus far named, but which is connected with the drainage systems of adjoining counties. The case is not explicable by existing agents of erosion. . . . The drift in the vicinty of Cynthiana often exceeds fifty feet in depth, and the origin of the great excavation which has here been effected must be sought in the glacial epoch, or in preglacial times."

Dr. G. Frederick Wright, in a number of articles published at various times, makes mention of the Beech Flats and the surrounding topography. In his recent work "The Ice Age of North America," 1891, on page 333, he gives a map showing the relation of the Flats and the head waters of Baker's Fork of Brush creek to the surrounding drainage systems, and makes some generalizations, based upon the Flats and certain features of Brush creek, which bear upon the important theory of the "Ice Dam at Cincinnati." While the value of this theory is not brought in question here, yet our conclusions, with reference to the origin of the Flats, lead to the belief that this theory must rest on other proof than that furnished by this region.

It has been my privilege to visit this region on a number of occasions, and personal examination has been made of almost the entire area represented on the map, Plate II. This map is constructed from data obtained from the following : " Map of the Marietta and Cincinnati Railroad, prepared by M'Gee and Phillips, Geological Locations and Sections by Prof. E. B. Andrews "; " Highland County, Ohio," from " Ohio Geological Survey "; Report of 1870 ; " Map of Highland, Ross, and Pike Counties "; of the " Ohio Geological Survey " Vol. II ; " Pike and Road Map of Adams, Brown, Butler, Clarke, Clermont, Clinton, Darke, Fayette, Franklin, Green, Hamilton, Highlands, Madison, Miami, Pickaway, Pike, Preble, Ross, Scioto, and Warren Counties ", " Geological Map of Ohio by Edward Orton " to accompany Vol. VI, " Ohio Geological Survey "; " Geological Map of Ohio by Edward Orton" to accompany Vol. VII, 1894, of the " Ohio Geological Survey"; with others. The topographical characters indicated on the map have been as accurately located as possible by sketch maps and field notes. They were located on the sketch maps in the field work, in relation to the pikes and roads, but it was not thought best to enter such details on this map as were not essential to the explanation of the work.

In order to get the facts presented in a connected manner, the reader is invited to accompany the author on an ideal trip of investigation, which is, however, with but one or two slight deviations, almost the exact duplicate of one of the trips taken during the study of the region. The line of this proposed trip is indicated on the map. (Plate II.)

Starting from Bainbridge, Ross county, our first point of observation will be the high quarry hill A just south of the town. This hill capped with Waverly freestone rises 450 feet above Paint creek, to an elevation 1180 feet A.T. A view to the north across the valley of Paint shows the hills forming the north wall of Paint valley over a mile away and rising nearly to a level with the observer. To the east extends the very deep and broad valley of Paint on its way to the Scioto. Large drift deposits fill the valley along its northern side, often rising to 150-200 feet above the creek. There are numerous terraces in the valley, and the creek is undoubtedly 150 feet above the rock floor of the valley. Westward the valley is distinguishable as a very evident trough of preglacial origin as far as the junction of Paint and Rocky Fork. Beyond rises very rapidly the drift-buried tableland of northern Highland county; and the well defined preglacial valley of Paint seems to end, suggesting some interesting problems in that direction, for somewhere in that locality we must look for the preglacial channel of reversed Brush creek.

As the view to the south is cut off by timber, we return to the pike and pass westward about one mile, where the view to the south shows a break in the high Waverly capped hills, and their place is taken by others of somewhat less altitude. From the pike along the creek the change in altitude would hardly be noticed; but as these hills are destitute of timber, they offer a prospect of an extended view to the south, and with this hope the ascent of the highest is undertaken. On reaching the summit at B, 190 feet above the river, and 990 feet A. T., it is found that this hill only concealed from the view on the pike still others just beyond, which rise just enough higher to obstruct the coveted view southward.

Our surprise is great, when almost on the summit of the next hill we find a ground-hog burrow, and the material revealed indicates glacial drift. The first thought is that this can not be drift at an elevation of 200 feet above the creek, and on the south side of Paint valley. A glance southward shows a comparatively level plain and not a rolling hill country as might have been expected.

This then is the extreme northern edge of the Beech Flats. What had concealed the real nature of these drift hills was their steep slopes and sharp summits. It seems almost inconceivable that these till deposits could have retained such steep gradients for such a length of time, yet it is quite evident that the high angles are the original slopes of the moraine and are not due to the subsequent erosion. There are many deep gullies and ravines cut by present agencies which reveal the true till structure of the deposits, but these can be readily distinguished from the older forms, although in both cases the slopes are so great that it is almost impossible to climb them. The surface is sparsely strewn with erratics. One of fine grained trap was estimated to weigh a hundred tons. There was also found a jasper conglomerate about the size of a man's head. Many observations in the immediate vicinity of B gave a mean elevation of 200 feet above Paint creek, 1000 feet A. T. The maximum elevation recorded was 250 feet above Paint, 1050 feet A. T.

The most conspicuous object in view from B is a high treeless hill to the west, which is located on the Giffen farm and which I have called Peach Orchard hill. The eastern exposure of the hill is very steep, and is much cut up by gullies which show very beautifully the contact line between the drift and the rock soil. The thin covering of boulder clay is pushed up the side of the hill at least 70 feet above the mean level of the Flats.

From the top of the hill, C, the prospect is grand and is well worth the climb. At an elevation of 485 feet above the creek, 1285 A. T., with an unobstructed view in every direction, it can not but enthuse the observer. The topography is spread out for inspection like a huge relief map, which it really is. Northward the view is similar to that from A as is also the view to eastward,—with this difference, station A is over two miles away with a broad and deep valley intervening. Following with the eye the outlines of this valley, as indicated by the long lines of hills, it is seen to extend many miles to the south, and within its rock-bound walls 250 feet below lies a tract of country, the Beech Flats, which never seem so flat as when viewed from this elevation. To the west extends the long ridge of sandstone hills which form the south wall of Paint valley. Just south of this is another ridge running nearly parallel with it with quite a wide valley between. As it is not possible to determine all the characters of this valley and its westward extension, we descend into it and proceed westward along the dirt road which runs along its northern side.

At D a spur extends southward from the Paint creek ridge, and the valley is much narrower here. It again widens to the westward. The small stream which drains it flows along the southern side and reveals the rock for much of the way to Rocky Fork at Barrett's Mills. At this point the valley seems to end, with the western ends of the two parallel ridges standing out in bold relief, with no visible counterparts on the westward side of Rocky Fork, which has here cut its way through the drift and developed its deep and picturesque gorge in the limestone.

From Barrett's Mills the journey leads along the range of hills bordering Rocky Fork. This ridge is crossed with the expectation of gaining Brush creek valley.

At E a small stream is crossed which is flowing westward instead of southward. It is at once recognized as not being Brush creek, and so is examined more closely and is found to flow into Rocky Fork between two high sandstone hills and in a rock gorge with vertical walls. This gorge is 75—100 feet wide and is clearly seen to be deeply filled with drift. It is very apparent that the gorge is not the work of the present stream, but that the latter is running, at least at the upper end of the gorge, much above the rock at an elevation of 940 feet A. T.

The next objective point is F, a very high cleared hill south of the village of Carmel. This hill reaches an elevation of 360 feet above Carmel, which is given as 939 feet A. T. From F the view is as grand and extensive as from Peach Orchard hill at C. The points most interesting in this study are the broad valley extending to the northeast to Cynthiana and filled with an arm of the Flats, and another broad valley very similar to the last, but stretching off to the southeast. The view to the northeast reaches to the horizon along a continuous valley. The view along the valley to the southeast is terminated in five or six miles by a line of hills, which are later found to be the hills forming the eastern wall of the Brush creek valley.

After observing a few land marks that will aid the recognition of our point of view, we descend into the valley again and traverse its rolling surface to Cynthiana. Here the drift shows a mean elevation of 200 feet above Paint creek, 1000 feet, A. T. Ascending a hill, G, just south of the village, our landmark at C is easily located, and it at once becomes evident that the axis of the valley, observed from C, passes east of Cynthiana and farther to the southwest.

Following along this valley, with its drift surface much cut up into hills and valleys (yet when on top of the hills this surface appears quite even) we reach a point H, near a large iron bridge across Brush creek, on the pike from Sinking Springs to Carmel. Here is observed a side valley entering from the northwest. Crossing the bridge and proceeding along the pike to a high hill of till at I, 191 feet above Paint creek, 85 feet above the water in Brush creek at the bridge, the landmarks at F are visible, and then is understood the reason why this valley when viewed from F, appeared to be closed at its southeastern end.

As we pass down the valley of Brush creek, it is noticed that while the creek is running in a large valley in the drift-filling it is nevertheless flowing with a very sluggish current. The high sandstone and slate hills are closer together and the rock valley is much narrowed. One looks in vain for a gap in the hills which will indicate the position of the exit of the creek. The only apparent opening is in the direction from which we have come.

We proceed to K along the bed of the creek which now* contains no water, except in the deep holes and is a muddy and sandy channel. Here is a small stream entering from the western continuation of the old valley. Here also are found evidences of a buried rock ravine which occupies a position more central to the main part of the old valley. This old ravine is filled with till and its walls but thinly covered. Its position is shown by a meander of the channel of Brush creek and also by a side ravine of recent erosion, which crosses it. The small tributary to Brush creek is fed by a number of springs and flows along the contact line of the drift and rock, along the southern side of the old valley, its former ravine near the center of the valley being filled with the drift.

Standing at the mouth of this small stream, not 500 yards from the place where the channel of Brush creek is transformed into a narrow and deep gorge, one unfamiliar with the facts would find no marks to indicate the location of the exit of Brush creek from this apparent basin, so skillfully has nature concealed the facts by topographical features, and a luxuriant growth of underbrush along the stream, gradually merging into the forest of the mountains sides.

Shults' mountain, L, presents the most favorable opportunity for a comprehensive view. From its summit, 440 feet above the waters

---

*August, 1895 ; an extremely dry season.

of Brush creek and 1325 feet A. T., the broad plain of the Beech Flats stretches away to the northeast and the horizon is formed by the hills forming the north wall of Paint creek, opposite Bainbridge. To the north at the foot of the mountain, lies the short westward extension of this valley beyond the exit of Brush creek. In the distance are visible the tops of the numerous ridges shown on the map. The appearance being that of a very hilly country and also resembling the view to the east and southeast. So similar are these two views, that the conclusion is inevitable, that their topographical features are the results of similarly operating forces. The region to the east and southeast is beyond the limits of the ice and the natural inference is, that the region to the north has been so slightly modified, that the main features of its preglacial forms are preserved.

Fisher's mountain M stands out farther to the southwest, and its summit is reached at an elevation of 410 feet above Brush creek, 1295 A. T. To the south lies a broad expanse of low lying country in the vicinity of Sinking Springs. From this low region the general slope of the country rises rapidly across Adams county to and beyond the Ohio river. Into this inclined plain Brush creek has excavated a narrow and deep gorge. So narrow are the gorges of all the streams in this district that from the point of view chosen it is impossible to determine their courses. From this same depressed region the country level rapidly rises to the west to the water shed separating the waters of Brush creek and Rocky Fork from those of western Highland county. This region is sparsely covered with drift. The land also rises rapidly to the northwest to the table land and drift-covered region of northern Highland county.

The descent is now made in order to study the characters of the Brush creek gorge which lies, as shown from our maps, between Fisher's mountain and Fort Hill, but which was not visible from the summit of the mountain. After a very steep descent of 410 feet, the rough mountain road passes between two walls of limestone, evidently a great fissure, and emerges in the dry bed of Brush creek gorge. The slopes of Shults' and Fisher's mountains and Fort Hill have angles of about 35 degrees and where these plains of the mountain sides would intersect occurs the $\bigcup$ shaped gorge of the creek, with vertical walls of about 50 feet and the gorge about 100-200 feet wide. The bed of the creek is composed largely of limestone gravel with a small percentage of northern drift pebbles. In no part of the gorge examined in this cut between the mountains was bed rock shown in the channel. The conclusion is inevitable that the considerable volume of water in the creek above the gorge must find its way out through the gravel filling in the bed of the gorge. The gorge of this fork of Brush creek was examined at many points to the southward and everywhere was found a small percentage of northern pebbles. This would be expected as the head waters are in the great till deposits of Beech Flats.

An examination of the data obtained reveals the following relations. The Beech Flats is a large tract of level land lying at an elevation of 1000 feet A. T, and 200 feet above Paint creek, occupying a portion of the southwest corner of Ross county, a portion of the northwest corner of Pike county and a portion of the eastern edge of Highland county. This land consists of a great deposit of till, showing but a few slight marks of stratification, being in fact largely typical boulder clay. It is bounded on all sides, except at its northern edge, with high rock hills, capped with Waverly freestone which reach an average elevation of 1200 feet A. T. At its northern edge, where it borders the valley of Paint creek, is exhibited an average of 200 feet in the thickness of the deposit. This thickness decreases towards the south and southwest, due to a rise in the rock floor, while the top of the deposit remains at nearly the same level. The elevation of this surface is the same as that of the drift deposits on the north side of Paint valley and also of those to the west in the vicinity of Barrett's Mills and Carmel, both of which are beyond the ranges of hills enclosing the Flats. The surface of the Flats is much cut up into drainage channels, so that the roads and pikes which traverse it in every direction are by no means level, but on the other hand are quite hilly. These channels in many cases seem to be out of proportion to the streams that occupy them. The two principal ones being, the small stream that rises near Cynthiana and flows west into Rocky Fork, the other and larger being Baker's Fork of Brush creek, which also rises near Cynthiana and flows south past Fort Hill and Shults' mountain. In both cases the valleys these streams occupy in the drift seem too large to be easily accounted for by present forces, especially in view of the slight amount of change observed in the forms of the deposits next to Paint valley. Both these streams pass out of the district through rock gorges. The former at an elevation of 940 feet A. T. and the latter 885 feet A. T. In both cases also the gorges are now partially filled and the streams are not flowing through the gorges on rock floors. The former stream has a fall of about 75 feet in its source to its gorge, the latter, Brush creek, has a fall of about 121 feet from Cynthiana to its gorge. The gorge of the former at E is 140 feet above the base of the drift at Paint creek, and the gorge of Brush Creek is 85 feet above the base of the drift at Paint creek. If it were possible to remove the drift from these valleys it is evident that these streams would be reversed. The fact is more striking when it is remembered that the bottom of the drift at Paint creek is 150-200 feet above the preglacial bed of Paint valley.

The conclusions which are drawn from the above facts are, that prior to the advent of the ice, the present location of Beech Flats was represented by a valley with numerous smaller tributary valleys, all tributary to the valley of Paint Creek. The heads of these valleys were at D, E, N, O, P, R, T, and V. At all of these points were cols connecting with adjacent drainage basins. As the ice advanced southward, planing and filling, it made the great drift plain of northern Ross and Highland counties and of Pickaway, Fayette, Franklin and Madison and other counties to the northward as its comparatively level ground moraine. It reached across the preglacial valley of Paint creek west of Bainbridge and pushed a great tongue of ice into Beech Flats valley. As this tongue advanced into this valley it divided again and again sending fingers along the tributary valleys clear to their head waters. Under these ice fingers was deposited the drift of the Beech Flats as a ground moraine. The spur which first separated from the main stream of ice crossed the col at D and probably joined the main mass of the ice sheet beyond Barrett's Mills. The next spur passes up the valley west of Cynthiana. The next spur passed up the next tributary valley to M. The main axis of movement continued beyond K to O, as shown by the till and boulders beyond the exit of Brush creek.

Large volumes of water were formed from the melting of these ice masses, for however much of rigor is attributed to the climate of the glacial period to account for the ice age, yet it seems evident that the margin of the ice extended beyond the line of perpetual snow and mean annual temperature of 32° F. into a temperate climate. The limit of the extent of the ice was determined by its rate of marginal melting as opposed to its rate of flow and supply of material by precipitation. The waters formed in the Beech Flats valleys found two outlets. One taking the water from the ice mass in the valley west of Cynthiana developed the gorge by cutting down the col at E. The main volume of the water flowed over the col at P and developed the **Brush** creek gorge.   As climatic conditions prevailed and the ice began to recede from the heads of these valleys the volume of water which filled the outlet streams at E and P remained large and excavated large valleys in the ground moraine thus exposed and cut deep and wide gorges in the rocks at E. and P.   As soon however as recession had proceeded as far as Paint creek this new channel was taken by the glacial waters and the water in Brush creek was suddenly reduced from a considerable torrent to a small stream fed only by meteoric waters.   As the volume of the water was reduced suddenly there was no opportunity for the development of terraces in Baker's Fork of Brush creek.   Subsequent erosion has resulted in the partial filling of these gorges and the present state of movement is now un certain.

If the above conclusions are warranted the Beech Flats must then be considered as a portion of the great level topped ground moraine so extensive within the limits of the ice movement in the portions of the State to the north and west of the Flats.

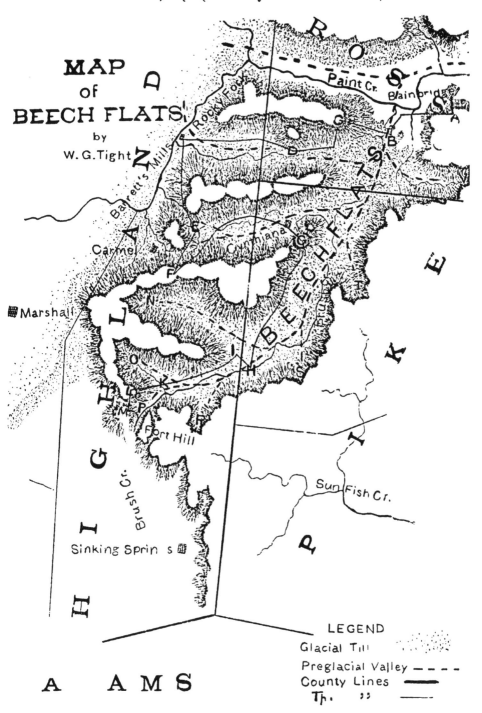

MAP
of
BEECH FLATS
by
W. G. Tight

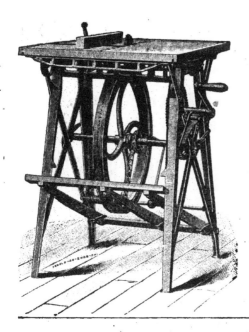
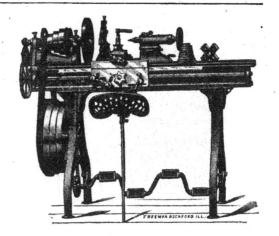

The Journal of Comp

AN

Illustrated Magazine Pɪ

DEVOTED TO

PART II.

With 11 Plates and 5 Figures.

EDITED BY

W. G. TIGHT, M.S.,

Department of Geology and Natural Hi

CONTENTS.

THE JOURNAL

OF

Comparative Neurology

A QUARTERLY PERIODICAL DEVOTED TO THE

Comparative Study of the Nervous System.

EDITED BY

C. L. HERRICK, PRESIDENT OF THE UNIVERSITY OF NEW MEXICO.

ASSOCIATED WITH

OLIVER S. STRONG, COLUMBIA UNIVERSITY,
C. JUDSON HERRICK, DENISON UNIVERSITY.

———

A technical journal containing original memoirs and abstracts, reviews and critical digests of the current literature of the nervous system in all its aspects. Fully illustrated by plates and figures in the text. Established in 1891 ; now in its seventh volume.

———

The Journal of Comparative Neurology

IS PUBLISHED BY THE EDITORS.

———

Subscription Price $3.50 per annum, post-paid to all countries in the postal union. Complete sets or single volumes of back numbers can be supplied at the regular subscription price. Single copies are sold separately, the price varying with the size of the number.

———

Address, C. JUDSON HERRICK, Managing Editor,

DENISON UNIVERSITY.   GRANVILLE, OHIO.

BULLETIN

OF THE

Scientific Laboratories

OF

DENISON UNIVERSITY.

VOLUME IX.

PART TWO.

EDITED BY

W. G. TIGHT, M.S.,

Department of Geology and Natural History.

GRANVILLE, OHIO, MARCH, 1897.

I.

WAX MODELING FROM MICROSCOPIC SECTIONS.

By W. E. Wells.

[Read before the Scientific Association at its Regular Meeting, Jan. 16, 1897.]

In the morphological and comparative study of organic structures the value of wax modeling cannot be over-estimated. In view of this fact, and owing to the apparent scarcity of literature on the subject, a brief outline of methods employed in the Biological Laboratories of Denison University, is here given.

Models may be divided into two classes: wax or clay models, which are moulded by hand, and wax and card-board models, which are constructed from microscopic sections.

Models of the first class are built up from observed form in gross dissection and will not be described in detail here.

Wax Modeling from Sections. This method consists, essentially, in constructing enlarged patterns of a series of microscopic sections, and from these patterns constructing a model which will represent the original unsectioned tissue, but on a greatly enlarged scale.

Wax. The material best suited for such modeling is ordinary beeswax. The beeswax of commerce is either bleached or unbleached. The unbleached, by virtue of its greater plastic property, is better suited for modeling. The least tendency toward brittleness becomes a source of trouble in the cutting process.

A medium sized model, measuring in its three dimensions, 6x3x2 inches, requires about one pound of wax, including necessary waste.

Commercial yellow wax may be had at a cost of from 40c. to 60c. a pound. Bausch & Lomb furnish a special modeling wax at 55c. a pound, in bulk, or in sheets of uniform thickness, at $1.12 a dozen.

Wax Sheets—Method for Casting. The wax sheets on which the sectional drawings are to be traced, must be of definite thickness, for if the length and breadth be magnified, the thickness must also be in direct proportion.

In rough modeling where general morphological relations only are sought, the thickness may be estimated, and sheets made accordingly.

In that case, the wax may be cast in shallow boxes, constructed from heavy oiled paper.

But in order to secure scientific accuracy, and usually this is desirable if not essential, it is necessary to have the wax sheets of known and exactly uniform thickness. In order to secure such sheets a casting box is necessary. A moulding box such as is used in the chalk plate engraving or stereotyping process answers every purpose. Such box may be improvized by using two perfectly smooth metal plates of suitable size. Between these plates, and on three sides, are placed narrow metal strips of the required thickness. The plates, having been previously warmed and oiled, are clamped together and the melted wax run in. The secret of obtaining good results lies in having the plates at just the right temperature. If too cool, the chilled wax will have a striated uneven surface. If too warm upon attempting to remove the sheets the wax will be found adhering to the plates. Care must be taken also to pour the wax in a steady stream, otherwise the sheet will contain air bubbles. It is necessary, moreover, to oil the plates before each casting. Vaseline is best suited for this purpose.

Tracing. The patterns are traced on the wax sheets by the aid of the camera lucida or the projection microscope. An ordinary lead-pencil or a porcelain stylus is used for the purpose of tracing. The latter is better followed by the eye on the dark wax background.

It must be remembered that the optical principles involved in the camera lucida require that the drawing surface shall be tilted toward the microscope twice as many degrees as the mirror is depressed below 45 degrees. Often when the magnification is high, or the object large, it becomes necessary to move the section in the field, which in turn necessitates a corresponding change in the position of the wax sheet. This is not only a tedious process, but necessarily introduces errors and is entirely obviated with the projection method. Often the desired magnification requires sheets too thin to work with. Such an obstacle may be met, if every other or every third section be drawn on sheets twice or three times the estimated thickness.

The latest development in wax modeling, as originated in our laboratories, is the process of wax tracing by means of the microscopic lantern projector.

A small movable screen is used on which to project the sections. By moving this screen in the focal plane, a graduated scale is made out to indicate different magnifications. On the screen is a frame which serves the purpose of a holder for the wax sheets. The magnification being determined upon, the wax sheets are cast with a corresponding thickness. The screen is placed at the proper distance as indicated by the scale, the images on the slides are projected in their serial order and the patterns traced on the wax sheets.

In case much detail is required, the contrast may be increased by covering the wax surfaces with thin white paper, and then tracing with a pencil, the pressure of the point being sufficient to transfer the outline to the wax beneath.

The advantages of this method are at once evident. In the first place it is labor saving. An entire series may be patterned within the space of an hour or two. Large models may be constructed without the extremely tedious process of moving the wax sheet while tracing, as is necessary with the camera lucida. And also, almost absolute accuracy is insured, which by the usual method is difficult to obtain, owing to the obscure field and possible lateral distortion of the camera lucida. It is only necessary to use an objective which will include all the field desired as the magnification depends upon the distance of the sheet from the lantern.

Cutting out Sectional Drawings. After the sections have been outlined on the wax sheets, they are to be cut out in serial order. This may be done, either with an ordinary pen knife, or better still, with an apparatus designed especially for this purpose. It consists of a wooden frame, between the two free ends of which a fine piano wire is stretched. The tension of the wire is varied by a thumb screw. A platform is attached to the lower arm, the wire being passed through a perforation in the center. As shown in Fig. 1.

When in use the apparatus is clamped to a table. The cutting is accomplished by simple pressure of the wax sheet against the wire. The smallest piano wire, with a diameter of from .037 to .035 of an inch, should be used. Experiments in heating the wire by electricity have so far proven unsatisfactory. For when the heat is increased to the proper degree, fusion of the cut edges takes place. All the sections can be cut at once, and then built up, or if only a small quantity of wax is available, the cutting and modeling may be carried on together section by section. Thus the waste may be used for casting additional sheets. If the first method is followed it is well to number the drawings to guard against any misplacement in the final reconstruction. When the sections have been fitted, they are cemented together by means of heat. The model is finally glazed over by being held for an instant over a hot flame.

Some models require mounting on a wooden base while others from the nature and position of the parts represented, show off to better advantage without any permanent base support.

Painting. To preserve and beautify the model, it should be painted. This also serves as an excellent means for differentiating the various parts. At Denison University a uniform series of colors are

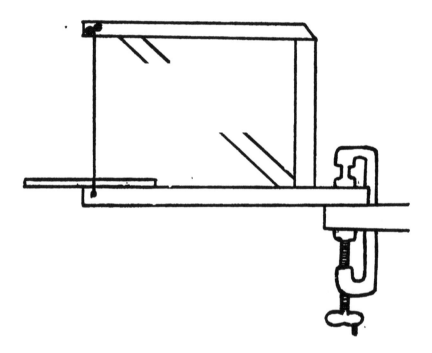

Fig. 1.

used to designate the different tissues. Thus, yellow always indicates nervous tissues, drab is used for epithelial tissue, red for glandular and black for bone and cartilage. White is generally used for painting permanent bases or supports on which the models are to be mounted. Each color should be laid on and allowed to dry separately. This prevents any running together of colors.

For black, turpentine asphaltum is unexcelled, for red and yellow tube paints answer very well. They require however, to be thinned out with a small quantity of turpentine. In fact all paints used for this purpose, should contain some dryer, such as turpentine or Japan dryer because the wax but poorly absorbs the oil.

Labeling. Labeling the different parts, adds greatly to the future value of the model. Any good mucilage answers the purpose of an adhesive, and is preferable to glue.

Modeling Instruments. Modeling instruments may be had from any biological supply house. However, those most necessary may be improvised without much trouble.

One of the most useful modeling tools is made by fitting a small spoon into a wooden handle. With this simple instrument almost the entire process of modeling may be carried out.

Another convenient tool is an ordinary long bladed knife. For rounding out concavities, a small iron bar with a spherical enlargement at one end, and fitted with a wooden handle, is a useful accessory.

Card-Board Modelling from Microscopic Sections. This method is a modification of the former. But serves to represent, in addition to form and relation of parts, the *histological elements* which enter into the tissues.

The process is essentially the same as that for wax modeling, except that the sectional patterns are traced on card-board instead of wax, and mounted on wires in a frame, instead of being fitted and cemented together.

After the sections have been cut out, the minute histological structures are sketched on each one, from a microscopical examination of the corresponding sections in the series.

Two or more wires are stretched within a light frame-work, in such a manner that the model is suspended in the required position. In each of the card-board sections, are two or more slots, through which the wires are to pass, when the sections are in position. The sections are separated from each other by small beads which are strung on the wire for that purpose.

By such an arrangement any section may be taken out of the model and its histological structure readily seen.

II.

ELECTRICAL WAVES IN LONG PARALLEL WIRES.

By A. D. COLE.

Read before the Am. Assoc. for the Advancement of Science, at Buffalo, Aug. 1896.

The experimental study to be described in this paper was undertaken as a preliminary to a research on the refractive index of certain liquids for electrical undulations as deduced from a measurement of the ratio of wave-length in the material under investigation to that in air. That research has been published in *Wiedemann's Annalen* (February, 1896) and in full abstract in the July number of the *Physical Review*, but so many facts not hitherto described were noted in the preliminary study that I have ventured to bring them before you in the present paper.

Stationary electrical waves were produced in two long wires according to Lecher's modifications of the original method of Hertz. The apparatus used is shown in Fig. 1.

I is an induction coil capable of giving a spark several centimeters long. Wires from its secondary terminals are joined to the two primary plates *P P'*, the latter being connected (except for a spark gap 2 to 4 mm long) by short rods terminated by brass balls.

The distance between the primary plates could be varied by sliding these rods in their support, and the resulting changes in the capacity and self-induction of the system controlled the oscillation period. The primary plates were 40 cm. square and from 3 to 10 cm. apart. From the oscillations set up in these by the discharges of the induction coil, oscillations were induced in two secondary plates, *s s'*, each 10 cm. square, placed a few centimeters in front of the primary plates. To the centre of each secondary plate a long wire was attached and these two wires, after approaching (as seen at *a a'*) to a distance of 8 cm. apart, stretched away horizontally and parallel a distance of about 4 meters. At every oscillation of the secondary plates a wave of electricity passed along each wire, was reflected back at its end and produced, by interference with new advancing waves, a system of stationary waves with alternating nodes and ventral segments, analogous to the stationary sound-waves in a vibrating stretched string. As in Lecher's experiments, a small Geissler tube without electrodes, placed between the remote ends of the parallel wires, glowed brilliantly.

A short wire placed across the parallel wires in general caused the light to cease, but positions could be found such that the tube still continued to glow. Three such positions were found with the apparatus used. These were separated by equal intervals and marked nodal positions, any interval giving the half wave-length of the undulations in the wires. The ends of the wires and of the secondary plates form ventral segments in the resonance system. H. Rubens[1] had succeeded

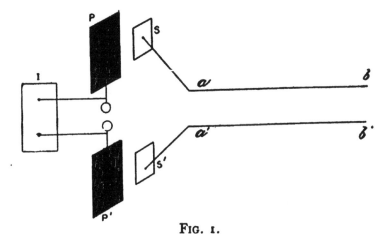

FIG. 1.

in measuring not only the length but the form and amplitude of such waves by use of the instrument devised by Paalzow and himself and named the "dynamo-bolometer." I used the same instrument employed by Rubens in his study of stationary waves in wires. Fig. 2 shows its construction. $R_1 R_2 R_3 R_4$ are the resistances of a balanced Wheatstone bridge. Two of these, $R_1 R_2$, are themselves balanced bridges, each of their four arms being a very fine iron wire about 10 cm long and of seven ohms resistance. Suppose the whole system balanced, and a weak but steady current supplied by the battery B. The galvanometer shows no deflection. Evidently if alternating currents produced by electrical oscillations enter by the wires ww' they will circulate only in the minor bridge R_2, but will disturb the balance of the main bridge by the heating effect in R_2. This disturbance, if

---

[1] H. Rubens, Wied. Annalen, Vol. XLII, p. 154, (1891.)

not too great, will be proportional to the heating effect producing it and this in turn, to the oscillations in the wires ww'. The galvanometer deflections become therefore a direct measure of the intensity of the electrical oscillations.[1]

To avoid disturbing the wave system in the wires, Rubens did not attach the wires w w^1 from the dynamo-bolometer directly to the points of the parallel wires to be investigated, but to little jars, made by placing around the wires bits of glass tube surrounded by shorter strips of metal foil as "outer coatings," the wires themselves forming the "inner coatings." With this apparatus, larger galvanometer deflections, ob-

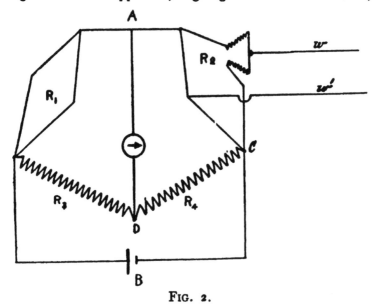

FIG. 2.

tained when the bridge wire is at a node, correspond to the glowing of the Geissler tube in Lecher's arrangement.

The wave distribution along the parallel wires was studied as follows: The Leyden jars were placed over the ends of the wires, a bridge of wire placed in a definite position across the wires (as shown by a tape measure stretched below), current sent through the induction coil, and the galvanometer deflection produced by the heating of the dynamo-bolometer read. This was repeated three or four times for each bridge position, and bridge positions taken 5 to 10 cm apart over

1. Rubens and Paalzow, l. c, Vol. XXXVII, p. 529, (1889.)

the whole length of the wire. Many such series of observations were made, the conditions being varied somewhat to learn the effect of such variations. The results may be best presented in the form of curves, abscissæ representing positions of the bridge on the wire and ordinates the corresponding galvanometer deflections in millimetres.

Rubens had noted that the number of maxima in a curve is diminished by (a) increasing the distance between the primary plates, and (b) leaving a bridge at a maximum position and exploring with a second sliding bridge.

These facts are shown repeatedly in my curves. Thus in curve D (Plate I) there are six strong maxima and seven or eight weaker ones. Abscissae represent position of the wire bridge as measured by a tape measure divided to centimeters stretched between the parallel wires, and ordinates the corresponding galvanometer deflection in millimeters. In this experiment the primary plates were 4.5 cm. apart. By increasing this distance to 8 cm. the three strong maxima (separated from one another by intervals of 146 cm.) almost disappeared, the other three remaining.

Again, by leaving a bridge in the position corresponding to the maximum nearest the secondary plates and exploring with a second bridge, all but three of the 15 maxima practically disappeared. These three occupied positions 8, 154.5 and 303.5 cm., at practically equal distances apart and are shown in curve G. The maximum near the middle of the wires was especially sharp, the two others weaker by the same amount. The effect of a fixed bridge is thus to reduce a somewhat complicated wave system to a simple one.

The maxima obtained were so improved in both strength and sharpness by the use of a second bridge that two very careful determinations of their position were made to get an idea of the degree of precision allowed by this method of calculating the half wave-length. The results are as follows :

	Maxima.			Intervals.		Mean.
First experiment,	25.3	188.5	351.4	163.2	162.9	163.1
Second "	24.8	188 4	351.5	163.6	163.1	163.4

The nodal positions seemed to be capable of being located to within a fourth of a centimeter, and on repeating the experiment the same results could be expected within a half centimeter.

Still greater steadiness and definiteness were secured in some later experiments by the use of three or even four fixed bridges.

No change has been attributed to a change of the distance between primary and secondary plates by other experimenters, so far as
I am aware; but a comparison of several of my curves seems to indicate quite clearly that diminishing this distance increases the complex-
ity of the curve.

It is worth noting that in each of these two and three following
careful determinations, the internodal space nearest the secondary
plates was about 5 mm longer than the other. This also has not been
elsewhere noted, so far as I know.

As it was my plan to estimate the refractive index of a number of
liquids by surrounding a portion of two parallel stretched wires by the
liquid under investigation, it seemed desirable to use the wires much
nearer together than other experimenters or I myself had done before, in
order to avoid the necessity of using a larger amount of liquid than I
could readily obtain. I accordingly set up my apparatus again, with
the wires only three cm apart. By this change two distinct results were
produced. In the first place, the deflections produced corresponding
to the maxima positions on the wire were considerably reduced in
amount, viz: to about one third of the value before obtained. This
was easily provided for. As the galvanometer used had been adjusted
to only moderate sensitiveness, a new adjustment of it gave deflections
sufficiently large. The other change noticed by bringing the wires
nearer was a shortening of the interval between nodal positions. To
make sure of this result and to measure the amount of the change,
three complete determinations of the three maxima were made, readings being made at each centimetre on the wire near the maxima positions. The results are as follows:

				Means.	Difference.	Previous.
26.6	26.7	.	26.5	26 6		
187.5	187.5		187.5	187.5	160.9	163.5
347.4	347.7		347.5	347.5	160.0	163.0

The three determinations agree so well that there can be little
doubt of a shortening of the internodal interval by about 3 cm or 1.5
per cent. But the true change is more than this, for the bridges themselves form a part of the resonance system, and two-thirds of their
length should be added to the apparent internodal distance to get the
true half wave-length. The bridges were 12 cm long in the first instance, and 3 in the second. This correction gave 171.2 as the half

wave-length for wires 8 cm apart, and 162.5 for wires 3 cm apart, a change of 9 cm, or more than 5 per cent.

It became desirable next to study the effect of surrounding a portion of the parallel wires by a containing vessel such as would be suitable for holding a liquid. I used a covered trough of zinc, 100 x 10 x 10 cm, with the wires passing centrally through rubber stoppers in the ends, one end being made to coincide with the centre maximum.

The internodal spaces, which had been equal before, were now different, that which included the metal box being shortened 3.3 cm or about 2 per cent.

The result is of the sort that we might expect, as the proximity of the metal box would naturally increase the capacity of a given length of the wires in that neighborhood, rendering a shorter length necessary.

Later experiments, in which there were two internodal spaces before the box was reached, developed the fact that the influence of the box upon the half-wave external, but adjacent to it, was considerable, since this half-wave was invariably 5 cm shorter than the one remote from the box.

When the box was filled with distilled water only a small, constant deflection of the galvanometer was obtained, and this was shown to be due partly to a direct magnetic effect of the induction coil upon the galvanometer and partly to current induced in the bolometer wires from the wires connecting the induction coil with the storage battery.

Up to this time the little "Leyden jars" had always been placed at the remote end of the parallel wires. The jars were now removed, and still smaller ones of the same sort placed at the ventral segments of the two external half waves. Although these had very small outer coatings—consisting of cylinders of copper foil 5 mm in diameter and 5 mm long—it was found that their capacity could not be neglected, but was equal to 2.5 cm wire-length. This appeared from the fact that the internodal space containing the "Leyden jar" was shortened 2.5 cm, the other remaining the same as before.

Still smaller jars were next made of a single turn of very fine wire about a glass tube 5 mm in diameter. These were found to make no appreciable change in the position of the maxima, and still, when placed at a ventral segment and connected to the dynamo-bolometer, sufficiently large deflections (100 mm) were obtained.

Very fine resonance systems were obtained in the external portion of the parallel wires in those experiments where a node was forced at the box end,—better than had been obtained with the parallel wires alone. This result is not attributed to the presence of the box, but rather to the use of the Leyden jars at intermediate ventrals segments instead of at the ends and to the readjustment of bridges already placed whenever the capacity of the system was changed by adding a new bridge.

Thus in one experiment with bridge fixed at box end 255, nodes were located as follows:

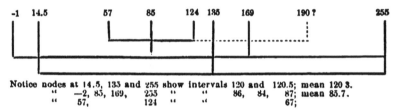

Notice nodes at 14.5, 135 and 255 show intervals 120 and 120.5; mean 120 3.
  "   —2, 85, 169,   255 "   "   86,   84,   87; mean 85.7.
  "   57,   124 "   "   67;

Thus the 9 nodal positions fall into three groups, showing intervals of 12.3, 85.7 and 67 respectively.

$$\text{Now } 120.3 \times 2\tfrac{1}{2} = 301$$
$$85.7 \times 3\tfrac{1}{2} = 300$$
$$\text{and } 67 \times 4\tfrac{1}{2} = 302$$

i. e., according as the nodal position occupied belonged to group 1, 2 or 3, the vibrating system consisted of $2\tfrac{1}{2}$, $3\tfrac{1}{2}$ or $4\tfrac{1}{2}$ half waves, the system of wires and end plates being equivalent to 301 cm of straight parallel wires.

These results gave a simple means of calculating the parallel-wire equivalent of the secondary plates and their connecting wires.   As the parallel wires began at −8 of the scale, the first node (−2) is 5 cm from one end, but the whole part of the system beyond this node $= \dfrac{86}{2}$ or 43 minus 5 = 38 as the equivalent in cm of the parallel wires for the capacity of each plate (6 x 6 cm) and of the 11 cm of wire leading from it to the parallel wires.

From another system of nodes we get in the same way the same result. Thus :

$$\dfrac{120}{2} = 60 \text{ minus } (7+14\ 5) = 38.5 \text{ or practically the same as before.}$$

From another experiment with a system of very different length of parallel wires, the wire equivalent of the same plate came out 39.2 and 40 ; mean, 39.6.

The distribution of energy in the internodal spaces was next investigated. Bridges were placed at two nodal positions external to the box and the Leyden jars moved from one end of the intervening half wave to the other at intervals of 10 cm. In each position four readings of the galvanometer were made and the mean values taken as ordinates for a curve whose abscissae were bridge positions, gave directly the distribution of electrical energy in the wires. For positions between successive nodes a very smooth and regular curve was obtained, which differs but little from a sinusoid. This is shown as curve N, Plate II.

When a bridge was placed at the box end, to force a node there, filling the box with distilled water left the wave system in the air-surrounded portion of the wires unchanged, but only slight and unsatisfactory traces of a wave system could be detected within or beyond the box, whether by Leyden jars placed at the end, or by those placed on the wires within the liquid. E. Cohn[1] had detected and measured the wave-length of resonance waves of this sort in water, and deduced therefrom the specific inductive capacity of water for long electrical waves. Of course the introduction of liquid, by the change in capacity, might be expected to destroy the resonance within and beyond the box, but it was hoped that by careful adjustment of a variable capacity at the end, the capacity of this part of the system might be increased exactly to some simple multiple of its former value and resonance thus restored.

These hopes proved delusive. The capacity at the ends was gradually raised by different means through wide limits, but such changes seemed to make no difference whatever in the resonance system within the liquid. This attempt was therefore abandoned, and improvement sought along the following lines: (a) securing great galvanometer sensitiveness, (b) purity in the liquids used, (c) using vessels of such materials that the liquids used would have no action upon them. Zinc had been used in earlier experiments. Now glass and glass coated with pure silver, alone were used. (d) Taking readings at very short intervals along the wire.

By the use of these various precautions well defined maxima were obtained, both with water and alcohol. (See Curve Q, Plate II). The maxima were less sharp and far weaker however in the part of the wires surrounded by water than in that surrounded by air. Since the

---

[1] E. Cohn, Wied. Annalen, Vol. XLV, p. 370 (1892.)

wave-length is much shorter in the liquid, (only one-ninth as much in water as in air), it was possible to obtain as many as four nodal points within a vessel 78 cm. long. The details of this work with liquids—water and alcohol—are given in the recent paper in *Wiedemann's Annalen* before referred to.

Since the above work was done Drude has published a paper[1] on the same subject giving preference to the exciter described by Blondlot over that of Lecher. I have accordingly made and used the Blondlot exciter shown in Fig. 3.

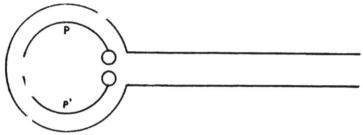

FIG. 3.

In this there are no plates, but the inner ends of the long parallel wires are joined by the circularly-curved wire of the figure. Within this wire and separated from it by only a few millimeters is the pair of primary exciters, P, P', bent to form arcs of a circle concentric with the curve of the surrounding wire, and carrying zinc balls for a spark gap at the inner ends. These exciters are connected directly to the secondary terminals of an induction coil. This form of exciter gave satisfactory results, but I have done too little work with it to make comparison with that of Lecher. I have also made successful use of the suggestion of Drude to use a Righi resonator, *i. e.* a strip of mirror amalgam with a narrow slit cut across the middle, as a means of locating nodes and loops in parallel wires. In a later apparatus of the Lecher type I have used balls of zinc instead of brass for the spark gap.

Neither Rubens nor Cohn seem to have found it necessary to protect the wires leading to the bolometer from electro-magnetic disturbances, but I have found it important. This was accomplished by enclosing them in long glass tubes which were in turn surrounded by a brass tube.

---

[1] P. Drude, Wied. Annalen, Vol. LV, p. 633, (1895.)

Curve G. Two Bridges.

Curve D. One Bridge.

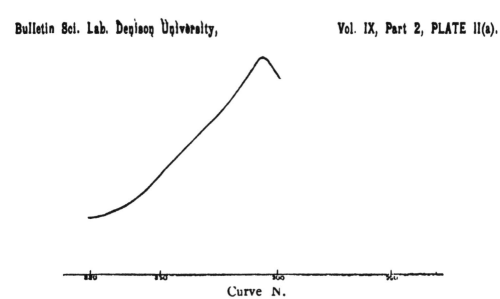

Curve N.

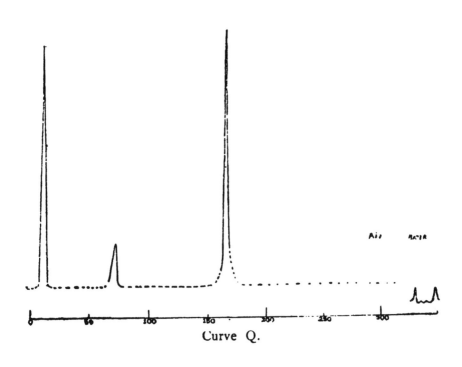

Curve Q.

To recall briefly the leading points:

1. Nodes could be located on the parallel wires to within $\frac{1}{500}$ of the internodal interval.

2. Resonance systems are injured in sharpness by placing the secondary plates very near the primary.

3. The effect of diminishing the distance between the parallel wires (from 8 to 3 cm.) is to (a) reduce the energy of the stationary waves to one-third of their former value and (b) to shorten the internodal interval.

4. The wire-equivalent of the capacity of the secondary plates is practically the same, when the wave length is considerably changed. This does not harmonize with the experience of Lecher.

5. If the parallel wires are made to pass through a vessel of water 70 cm. long, capacity added to the remote ends does not affect the resonance system in the wires, either within or without the vessel.

Most of the experimental work described above was done at the University of Berlin under the direction of Dr. H. Rubens, but the later portion in the laboratories of Denison University, Ohio.

III.

CHANGES IN DRAINAGE IN SOUTHERN OHIO.

By FRANK LEVERETT.

With One Plate.

In connection with a study of the Ohio Valley and its tributaries, carried on for the U. S. Geological Survey, for a period of several months the past year (1896), I was so fortunate as to discover an abandoned valley departing from the present Ohio at Wheelersburg, Ohio, about ten miles above Portsmouth, and passing northward in a somewhat winding course to the Scioto River opposite the city of Waverly. (See map, Plate II). The valley is fully a mile, and perhaps 1½ miles in average width, and is cut to a depth of nearly 300 feet below the general level of the bordering upland, and to within about 150 feet of the present level of the Ohio. A part of this valley was long since noted by Dr. Edward Orton as the channel of a large stream, but its connection with the Ohio was not worked out. (Geology of Ohio, Vol. II, 1874, pp. 611-12).

This valley is plainly the channel of a north flowing stream, and carried the Great Kanawha and Big Sandy drainage, as well as that of several smaller tributaries of the Ohio, together with a small section of the present Ohio Valley. Evidence of the northward flow is found both in the slope of the rock floor and in the character of the river debris. A series of careful aneroid determinations indicate that the rock floor falls 25 feet in passing from Wheelersburg to Waverly, a distance by the windings of the valley of about 30 miles.

On the rock floor is a deposit of well rounded pebbles and larger stones such as characterize river bottoms. These deposits though now covered with 25 to 50 feet of silt are exposed by modern ravines which show them to be usually several feet in depth. The stones range in size from a foot or more in diameter downward to fine pebbles.

The significant feature in connection with this river debris is the kind of rocks. They are very largely made up of quartzite and pebbles formed from vein quartz, such as are abundant in the terraces of the Kanawha System of West Virginia. The fact that such stones are sparingly represented in the form of boulders imbedded in the Coal Measures of southern Ohio, makes it necessary to determine whether they are of local or of distant derivation. The rarity of these boulders in the Coal Measures, however, is such as to render it improbable that the large number of quartzites lodged in the abandonded valley could have been derived from the immediate vicinity. It seems far more probable that they were brought by the Kanawha System of drainage from extensive outcrops of such rocks on its head waters, notably along New River.

This abandoned valley forms a natural continuation of the old Kanawha System, which, as shown some years ago by Prof. I. C. White (Appendix to Wright's Glacial Boundary in Ohio. Western Reserves Hist. Socy. Cleveland, Ohio, 1884, page 84), and discussed more fully later by Prof. G. F. Wright (Bulletin U. S. Geological Survey No. 58, 1890, pp. 86-88), discharged westward from near St. Albans, W. Va , through the abandoned channel known as " Teases Valley," to Huntington, W. Va., and thence down the present Ohio. There is a slight departure from the present course just below the mouth of the Big Sandy, near Ashland, Ky., where for a few miles it passed through a broad channel lying just south of the present south bluff. This channel back of Ashland was long since noted by Mr. Lyon of the Kentucky survey, and afterwards described by Prof. E. B. Andrews of the Ohio Survey. (Geology of Ohio, Vol. II, 1874, p. 441). The course of these abandoned channels may be seen on the accompanying map, Plate II.

The old rock floor of Teases Valley stands about 650 feet above tide, or very nearly 150 feet above the present Ohio at Huntington. The rock floor of the old channel, as preserved in numerous remnants between Huntington and Wheelersburg, shows about the same rate of descent as the present stream. At Wheelersburg it stands about 625 feet above tide. Following the abandoned valley north the rock floor descends to about 600 feet at the point where it joins the Scioto, opposite the city of Waverly. Teases Valley, and also the channel back of Ashland, and the remnants along the border of the present Ohio, all carry a deposit of rolled stones made up largely of quartzite, and similar in every way to the deposits of the abandoned valley leading north from Wheelersburg.

In the portion of southern Ohio east of the Scioto, from the present Ohio northward at least to the Hocking, the streams now directly tributary to the Ohio, have in several instances been greatly enlarged at the expense of streams tributary to the southern end of the Scioto Basin. A reference to the accompanying map will show that the present drainage systems are very abnormal. There are suggestions of still greater changes not yet worked out to a demonstration. The Little Kanawha, with also several smaller southern tributaries of the Ohio, and a considerable portion of the Ohio itself above Huntington, may have discharged somewhat directly westward to the Scioto Basin across southern Ohio, instead of taking the roundabout course by Wheelersburg. I have only examined a part of the district which would be traversed by such a line or lines of discharge, so cannot speak with the confidence that I do of the other changes noted, but the following statements may be made.

It is thought that this westward flowing system may have received drainage lines from nearly as far southwest as Teases Valley, and that in developing the present Ohio System a col may have been crossed by the Kanawha near Winfield a few miles north of St. Albans, and by the Ohio only a few miles above Huntington.

It is evident that the great part of the present drainage basin of Symmes Creek was once tributary to the Scioto through Salt Creek, there being a broad abandoned channel leading north past Camba, connecting its head waters with Salt Creek, near the city of Jackson. (See map, Plate II.)

A probable change is to be found in the lower Scioto Valley. This may have once received the several small streams which flow in a northeasterly course and enter the Ohio nearly opposite the mouth of the Scioto, and then have carried the waters north to join the old Kanawha at Waverly. This small drainage basin would include also that portion of the Ohio (reversed) between Buena Vista and Portsmouth. There is a bare possibility, however, that the Kanawha System turned south at Waverly, and followed down the Scioto and Ohio.

The northern part of the Brush Creek drainage basin certainly was once tributary to the Scioto, as indicated by Messrs. Tight and Fowke in a former bulletin, and it is possible, as suggested by Professor Tight, that the entire Brush Creek drainage basin once had northward discharge into the Scioto Basin, carrying with it a small section of the Ohio between Vanceburg, Ky., and Manchester, Ohio.

Concerning the direction of discharge for the old Kanawha System from the south end of the Scioto Basin, but little is known. The

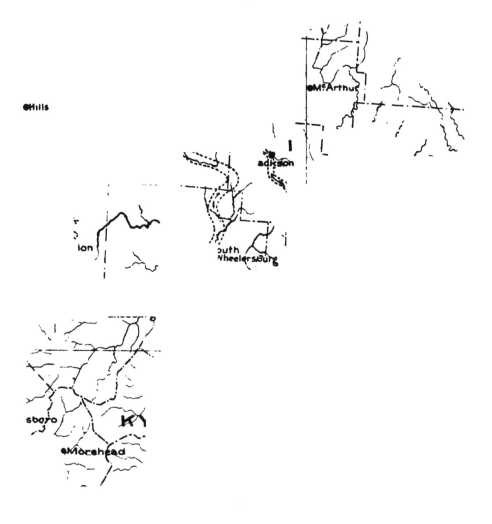

Scale

EXPLANATION OF MAP.

The course of the old Kanawha System from St. Albans, W. Va., to Waverly, Ohio, is indicated by dotted lines representing the breadth of its channel. The northward discharge from the Symmes Creek drainage basin is similarly shown.

The position of cols is indicated by a cross bar. In cases where some doubt is felt concerning the col a query mark (?) is affixed.

The vagueness of the glacial boundary in this region is such as to render it difficult to determine with accuracy. It should be noted that the boundary here represented lies a few miles outside the lines which Prof. G. F. Wright has published. The departure however seldom reaches 10 miles.

heavy deposits of drift in central, northern, and western Ohio, render it very difficult to trace a northward line of discharge. There are at least four possible courses to be examined: 1. Southward, down the Scioto from Waverly to the Ohio, and thence down the Ohio; 2. Northward along the axis of the Ohio Basin to Lake Erie; 3. Northwestward across western Ohio, along one of the several deep valleys brought to light in that region by the oil and gas wells; 4. Northeastward past the Licking Reservoir, and along the old valley (brought to notice by Professor Tight in a former Bulletin), to the Muskingum at Dresden, thence northward along or near the present valleys of the Muskingum, Tuscarawas, and Cuyahoga to Lake Erie at Cleveland. I have thus far been unable to rule out any one of these lines as an impossible one, or to reach any satisfactory conclusion concerning the probabilities of the case.

Of the three most influential factors likely to have been potent in causing the changes of drainage in this region, uplift, stream piracy, and glaciation, the last mentioned is the only one known to have been an effective cause. That it is the only important factor is however by no means certain. The question of the cause, or causes, of the changes should therefore be left open.

Washington, D. C., January 18, 1897.

IV.

SOME PREGLACIAL DRAINAGE FEATURES OF SOUTHERN OHIO.

With Plates.

By W. G. TIGHT.

[This paper was presented to the Ohio Academy of Science at its winter meeting, Dec. 26 and 27, 1896, under the title of "The Big Kanawha Drainage."]

CONTENTS.

1. Introduction.
2. Some Features of the Ohio River Valley.
3. The Big Sandy Valley of Kentucky where it enters the Ohio.
4. The California Valley from Sciotoville to Waverly.
5. The Symmes Creek and Salt Creek Valleys.
6. Correlation of the Drainage and Topographical Features.
7. A Pseudocol.

I. INTRODUCTION.

In the Spring of 1895 it was my pleasure to spend some time in field work in southern Ohio and parts of Kentuckey and West Virginia. The results of that work were at once prepared for publication but were withheld on account of the fact that it did not seem that sufficient work had been done in the field in a certain locality to complete the data in hand and arrive at a satisfactory conclusion. The desired data have recently been obtained and inserted in the original manuscript in the proper location with the necessary alterations.

Lest there should be some misunderstanding it seems best to state here certain facts with reference to the investigation of this region in question. As will appear later in this paper in their proper order the following facts were determined during the investigations in 1895. (1) That there was undoubtedly an old col in the Ohio Valley just above Portsmouth. (2) That there was an old high level valley running northward from the Ohio Valley between Sciotoville and Wheelersburg. (3) That there was an old valley opened eastward from the Scioto valley just north of Piketon and opposite Waverly. (4) And from Dr. Edward Orton's description in the Ohio Geological Survey, Vol. II, p. 611 that there was an ancient drainage feature in southeastern Pike County.

In the spring of 1896 Mr. Frank Leverett, of the United States Geological Survey, while engaged in field work in Ohio, called on me and I gave him the results of my work. I also stated to him at that time that it was my opinion, if as I suspected a continuous valley should be found to extend from Sciotoville northward past California Flats to the Scioto valley at Waverly, it must be connected with the " Flat Woods and Teazes Valley drainage and substantiate more fully the general drainage line indicated in my former article in this series, Vol. VIII, Part II, Plate V.

He at once planned to visit the region and after about two weeks in the field Mr. Leverett returned and reported that he had found it as expected and had traced the valley all the way from Wheelersburg to Waverly. Some months later it was possible for me to visit the region also and the data given in this article and shown in the plates and map were obtained during my visit. As the credit of first establishing this old valley belongs to Mr. Leverett I asked him to contribute to this series his interpretation of the region. He has complied and article III of this series and volume is from his hand. Whatever similarity the reader may find in the substance of these two articles or the plates presented with each should be taken as increased evidence of the truthfulness of the observations and conclusions as the two are of independent origin, except so far as stated above.

2. SOME FEATURES OF THE OHIO RIVER VALLEY.

The great trough of the Ohio river presents many problems of interest to the geologist and within a few years some new and startling theories have been advanced with reference to certain cycles of its development history. The truth or falseness of these theories remains to be proven by the accumulation of evidence for or against them.

In a former part of this series Vol. VIII, Part II, page 60, the writer expresses the belief that the Ohio river valley, in the portion along southern Ohio, is made up of modified parts of other preglacial drainage systems and its present position largely determined by the position of the lowest cols between the various elements.

The object in mind in the field work was to see if sufficient evidence could be found on which to locate these old cols. The task is by no means an easy one on account of the size of the valley and the vast amount of erosion which has taken place through the entire length of the valley. While the data here presented are not of a very exact nature from the fact that the time which it was possible to devote to the field work did not permit of detailed measurements, except in a few cases; yet it is hoped that they may be of value in suggesting fields for future investigation.

The section of the valley included in this study extends from several miles above Huntington, West Virginia, to Vanceburg, Kentucky. The points to be considered are principally the width of the valley, measured from the point of intersection of the flood plain or terrace filling with the base of the rock wall on one side to a corresponding point on the opposite side; The character of the slopes of the rock walls whether steep and precipitous or gently sloping; the stratigraphic relation and character of the rock as to its disintegrating properties; the elevation of the valley walls above rock floor and present water level; the presence of elevated rock and gravel terraces; and the character of tributary valleys. While it may be possible by such characters to locate the position of old eroded cols in the valley modifications of small streams to within a few hundred yards. (Article V.) One would be fortunate to locate by detailed study the cols of such a valley as the Ohio within a few miles, unless the characters were very evident. The section of the Ohio from Vanceburg to some miles below Manchester will be considered in an other article, now in preparation, on the Brush Creek drainage. Suffice it here to say that in the vicinity of Vanceburg and for some miles below the valley of the Ohio is relatively narrow. In places it was estimated to be less than a mile wide. The bordering hills are high and precipitous. The rocks are exposed in vertical cliffs with sharp upper angles; although composed of shales, limestones and sandstones which are not especially resistant to disintegration.

Passing up the river towards Portsmouth the valley grows wider and the bordering hills while fully as high do not have as abrupt faces towards the river front. Although the slopes are everywhere at high angles. About 4 miles below Portsmouth near Scioto Heights at the mouth of Rock Run there is a rock platform of considerable extent which appears very much as though it might be an old gradation plain, although no stream trash was certainly identified upon its surface which is about 130 feet above the river. Scioto Heights rises, with about a 45° slope, to 510 feet above low water in the Ohio. Here the Ohio Valley is estimated at one and a half to two miles wide. The valley widens perceptibly above the mouth of Kinniconick Creek which meets the Ohio from the south west flowing in the opposite direction to that of the Ohio. Passing up the Ohio toward Portsmouth the valley of the Ohio seems to be directly continuous with that of the Scioto. The Scioto valley at its mouth is some wider than the Ohio immediately below. If a stranger unfamiliar with the facts would not observe the volume of the waters in the Ohio above the mouth of the Scioto and the latter stream, but would base judgment on the form, size and directions of the valleys, the Scioto would be taken as the continuation of the Ohio Valley as it is seen approaching the junction of these streams both from the Scioto and lower Ohio valleys. The Ohio valley immediately above Portsmouth is scarcely a mile wide while bold cliffs of Waverly shales and sandstones face the stream on both sides of the valley.

A few miles further up the river the valley grows wider and receives a considerable tributary, Tygart's Creek, from the southwest. At this point also the valley bears to the northeast toward Sciotoville and in this direction the valley also increases in width.

At Sciotoville begins the great bend in the Ohio valley towards the south and southeast. Here also enters from the north the Little Scioto river which as will be shown later is a reversed stream with a deep valley cut out of a former drainage system.

Continuing up the Ohio valley almost due south to Greenup the valley is much wider. The bordering hills do not rise to the level of the cretaceous peneplain until some distance back from the immediate valley walls. The travelers on the river steamers do not seem to be so shut in but enjoy a more open and extended view or may do so if their eyes are open to the world around them.

The old drainage level valley known as " The Flat Woods " which runs parallel to the Ohio on the Kentucky side back of Ashland will not be described. The reader is referred to the literature on the subject as found in Kentucky Geological Survey, described by Mr. Lyon ; Ohio Geological Survey by E. B. Andrews Vol. II, Page 441; by Prof. G. F. Wright, Vol. V, Page 765.

At Catlettsburg the Ohio receives a large tributary from the south, the Big Sandy. The Ohio valley is perceptibly wider below the mouth of the Big Sandy than above it.

At Huntington the Ohio Valley is about one and three quarters miles wide. Here also enters from the north through a narrow deep valley the considerable stream of Symmes' Creek.

At Guyandotte the Guandotte river joins the Ohio and at this point also the Teazes valley meets the Ohio. For descriptions of the Teazes valley the reader is referred to Bulletin Geological Survey, No. 58, page 86, Prof. G. F. Wright, Ice Age of North America, page 339, same author.

Passing further up the river the valley rapidly grows narrower. The characters of the valley resemble those noted below Vanceburg and above Portsmouth. There seems to be nothing in the stratigraphy which would produce this narrowing of the valley and the presence of the bold cliffs of carboniferous rocks. Several miles above Guyandotte the width of the valley was estimated at less than a mile. Observations were not extended farther up the river.

The tributary valleys of the Ohio to which attention is especially directed are: The Teazes valley of West Virginia and the Flat Woods valley back of Ashland, Kentuckey, both of which have been described by other writers; The Big Sandy; The California valley and Symmes' Creek.

3. THE BIG SANDY VALLEY.

Our examination of the valley of the Big Sandy did not extend above five miles from Catlettsburg and the results of the study will be briefly stated. The general direction of the valley conforms to that of the Ohio from Sciotoville to Catlettsburg. A cross section would show very clearly that the present channel has been eroded out of a more elevated valley floor. The rock platforms left on both sides of the valley show the old drainage level to be [by aneroid] about 150 feet above the present level of the Ohio. The old gradation plain is in many places very prominent. The characters of this high level valley resemble those of the Teazes, Flat Woods, and California valleys. There is no doubt in the mind of the writer that the gradient of the Big Sandy has been recently increased thus producing rapid cutting in the old gradient plain.

4. THE CALIFORNIA VALLEY.

This is the name which Dr. Edward Orton gave to a portion of an old deserted drainage line which extends from Sciotoville to Waverly along the line indicated on the map Plate III and which will be described more in detail. Dr. Orton, in his report on Pike County, in the Ohio Geological Survey, Vol. II, page 611, says: " In the extreme northwestern and southeastern corners of the county, near Cynthiana[1] and California respectively, there are conspicuous examples of surface erosion that do not belong to either of the systems thus far named, but which are connected with the drainage systems of adjoining counties. Neither case, in fact, is explicable by existing agencies of erosion. The California valley, which is very broad and deep, is occupied by an insignificant stream that flows with a sluggish current upon the surface of the deep drift beds by which the valley is filled. The Drift in the vicinity of Cynthiana often exceeds fifteen feet in depth, and the origin of the great excavation which has here been effected must be sought in the glacial epoch, or in pre-glacial times."

Whether Dr. Orton recognized the continuation of this old drainage feature southward from California through Scioto county to the Ohio valley is not stated. But that he recognized the main features of the northern portion of the valley is very clear, so that the name which he gave to the part is here retained for the whole valley. In company with Mr. Wiltsee, of the department of geology, I examined this drainage line from the Ohio to the Scioto. The valley of the Ohio from Greenup to Wheelersburg continued northward would follow directly into the Ohio end of the California valley. North of Sciotoville and Wheelersburg the gradation plain of this valley lies about 150—175 feet [by aneroid] above the present Ohio. The rock floor gradually descends as the valley passes in a great sigmoid across Scioto county to California. Here it is estimated from well depths that the rock floor is about 100 feet below the surface. The valley next makes a great bend to the eastward into Jackson county and then westward to the Scioto. The southern portion of the valley floor has been much cut up by recent drainage lines but in many places the gradation plains are preserved and on the old valley floor was found river rubbish exactly similar to that of the Teazes and Flat Woods valleys. The little Scioto

---

[1] See description, Bulletin, Vol. IX, article III, this series.

river has worked out a deep and rather broad valley in this region but a glance at the map plate III will show that the Little Scioto does not follow entirely the line of the old valley. The Rocky Fork occupies the old valley for a considerable distance, than leaves it to join the main fork and together they gain the old valley again farther to the south. There has been such a vast amount of erosion along the lower portion of the Little Scioto (and considered in connection with other characters) as to make it very probable that some of the work was done by a larger stream than the present river.

The old valley floor lies about 300 to 350 feet below the hills which border it and in no case were cliffs observed on either side of the valley although often the slopes were quite steep. The similarity in the topographic forms of this valley to those of the Teazes and Flat Woods valleys and the portions of the Ohio connecting these elements is very marked, while the dissimilarity to the forms below Vanceburg and above Portsmouth and Guyandotte on the Ohio is as striking.

In the northern portion of the valley the width increases until it is even greater than that of the Scioto below Piketon. The rock floor is here deeply buried beneath a very compact, finely laminated river silt, good sections of which are revealed in many places by recent erosion.

The descent of the rock floor also continues to the northward. The present drift plain of the valley presents a high terrace like front along the Scioto valley.

5. SYMMES CREEK AND SALT CREEK.

Symmes creek rises near the center of Jackson county and flows southward through Gallia and Lawrence counties and joins the Ohio opposite Huntington, West Virginia. This valley was studied at only two points. Reference has already been made to the character of the valley at its junction with the Ohio where the stream runs in a very deep and narrow trough which it has cut out of the carboniferous rocks.

At the head waters near the city of Jackson the stream is in a broad and open valley which it shares with the head waters of the south fork of Salt creek which latter flows northwestward into the Scioto at the great bend in the southeast corner of Ross county. The old erosion valley at Jackson is over a mile wide and is cut some 200 feet into the table lands. The valley floor rises to the southward and the width of the valley decreases somewhat in the few miles that it was examined south of Jackson. To the northward the valley broadens out rapidly towards the Scioto. The fact that this valley was not produced by the action of its present stream was early recognized by Dr. Edward Orton for in his report of Ross county, Ohio Geological Survey Vol. II, page 642, he says: "East of the Scioto, and in the southeastern corner of the county, Salt creek flows in an old and deeply excavated valley."

Time did not permit the examination of this valley its entire length but it seems very probable that some connection may exist between this valley and the drainage basin of Raccoon creek. The location of the col on Symmes creek as shown on Plate III was made entirely from the map study and was not verified in the field so that it is only tentatively placed. The northward direction of all the tributaries and especially that of Sand creek is very suggestive of the fact of the reversal of drainage and was the principal factor considered in locating the col.

6. CORRELATION OF THE DRAINAGE AND TOPOGRAPHIC FEATURES.

The reconstruction of the old drainage lines of the region seems very plain. The main axis of drainage was the Big Kanawha, with its head waters, in New River, in the old land of the Blue Ridge, it crossed the low inland of the great Appallachian valley and the Allegheny plateau along the line of the Teazes valley to the Ohio at Guyandotte. Here it received a tributary from the north along the line of the Ohio which headed at near Millersport, also one from the south, the Guyandotte river. Its course then conformed to that of the Ohio to the mouth of the Big Sandy, thence it followed the Flat Woods valley to Greenup where it again conformed to the present Ohio to Sciotoville. At Greenup it received a tributary from the south, the Little Sandy. At Sciotoville it also received an other southern tributary, Tygarts creek, the lower portion of which conforms to a portion of the Ohio. The main stream continued northward through the California valley to Waverly where it received an other considerable tributary on its western side which was made up of the waters of Salt Lick creek and Kinniconick creek continued along their normal directions along the reversed Ohio and lower Scioto rivers. The great ridge which separates the waters of Kinniconick creek and Tygarts creek then continued northward right across the present Ohio just above Portsmouth into Scioto county and formed the water shed between the main stream and this last mentioned tributary at Waverly. The presence of this great drainage line along the California valley so close to the lower Scioto and this old divide easily accounts for the lack of any large tributary stream on the east side of the lower Scioto.

From Waverly the main axis continued northward along the present line of the Scioto into central Ohio. At the great bend of the valley in southeastern Ross county a considerable tributary was received along what I have called the old Jackson valley. As shown by Mr. Fowke (article II this series) another large tributary was received just above Chillicothe, the Paint creek drainage.

The course of this great drainage line from central Ohio may be somewhat doubtful but the writer has as yet no grounds for modifying the views expressed in his first article in this series, in which he states that the Preglacial Muskingum and reversed Scioto found a western outlet along the line roughly indicated on the map, Plate V, accompanying that article. The facts here presented and the conclusions drawn seem also in harmony with the views expressed that the Ohio river valley along southern Ohio has been developed, in very recent geologic time, from the adjacent parts of several older river systems by the cutting down of the old cols between these basins. The silting up of portions of the old channels during the back high water stage above low neighboring cols undoubtedly determined some of the important modifications. It would seem as though the filling of the northern end of the old California valley may have produced the deflection of the waters across the ridge between Sciotoville and Portsmouth. If however the waters rose high enough to occupy both courses, leaving the area included by the Ohio, California and Scioto valleys as an island it would be expected that the shortest course would be developed on account of the greater grade.

7. A PSEUDOCOL.

In determining drainage modifications it is necessary to locate with accuracy the position of the original col which was worn away in the development of the new system. To accomplish this many things are taken into consideration. The criteria will be quite different when the systems bear different relations to each other, notably when the new system is above or below the level of the old and when the new system is larger or smaller than the old. Without entering into a full discussion of this question I desire to call attention to a particular case which has come under my notice where most of the criteria seem to be fulfilled for the location of an eroded col and yet the peculiar topographic form has been produced in an entirely different manner.

In the case in which the new system carries a larger or smaller volume of water but has not as yet reached the depth of the old system, or in other words when the new system is larger or smaller but above the old, the location of the col is partially indicated by the broadening of the valley and the descent of the rock floor of the valley in opposite directions from the col. Other adjacent streams usually indicate also the position of the divide in which the col occurred. These conditions may exist where there has been a modification and reversal of

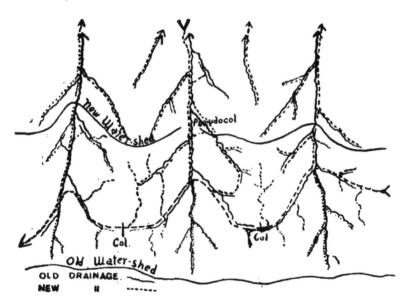

drainage without there ever having been a col at the point suggested by the data and the divide which at first would seem, from map study, to be a part of the old form is in reality a result of the combination of the two forms. The conditions would be produced when one large drainage line is developed transverse to another with a reversal of part of the old system and an increase in the volume of water (or lapse of considerable time). The relations are indicated diagramatically in the accompanying figure in which the old drainage is indicated by the con. tinuous wavy lines; the new drainage, by the dotted wavy lines.

The form at A resulting from this combination I have termed a pseudocol from the fact that the data so strongly indicates the existence of an eroded col that one is apt to be misled.   In the case of the true col not only the gradient of the rock floor of the valley descends from the location but also the general topographic surface descends in a direction at right angles to the true divide, while in case of the pseudo-col the general topographic surface forms a continuous plain across the new system and the apparent divide as indicated by adjacent drainage lines is the balance line between the two cycles and is evidently migrating rapidly from the new drainage axis.   Field illustrations of this peculiar topographic form will be given and discussed later.

Photo and Eng. by W. G. Tight.   A View of "California Flats" at A, Plate III.   Bordering Hills on the Left.

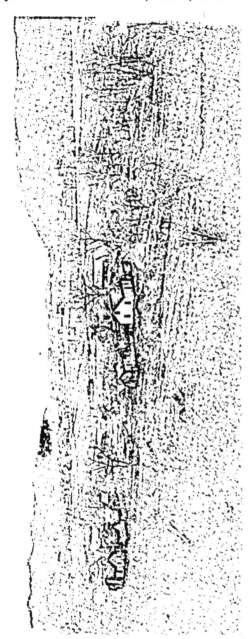

Photo and Eng. by W. G. Tight.

Looking up the Little Scioto Valley from B, Plate III. The valley floor of California Valley is seen in the middle distance and the old valley wall at the horizon.

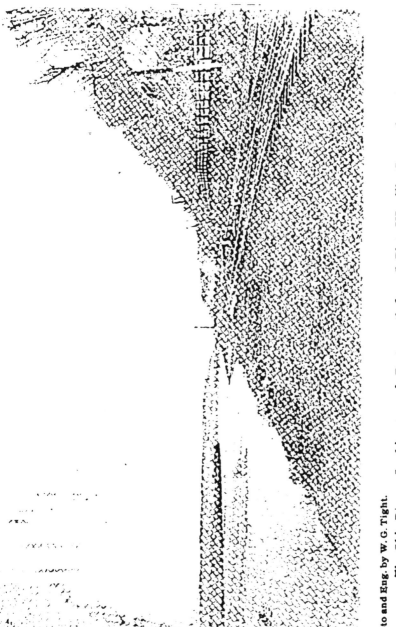

Photo and Eng. by W. G. Tight.

The Ohio River. Looking towards Portsmouth from C, Plate III. The Gorge character of the Valley is shown in the distance.

LEGEND
Scale 0   5   10 m.

County Lines ——

Streams ~~~~~

Valley Walls - - - - -

N

S

V.

A PREGLACIAL VALLEY IN FAIRFIELD COUNTY.

By W. G. TIGHT.

[Read Before the Ohio State Academy of Science, Dec. 26 and 27, 1896.]

My attention was first called by Dr. J. C. Hartzler of Newark to the existence of a preglacial drainage line east of Lancaster, in Fairfield county. In company with Dr. Hartzler and Prof. Richards, of the Newark High School, I visited the region in the Spring of 1896 and together we traced a portion of the old valley. A more extended examination of the region was made in the Fall in company with Mr. C. A. Wiltsee, of the department of Geology of Denison University.

The area under consideration and shown on the accompanying map, Plate IV, includes principally Berne and Rush Creek townships of Fairfield county, and Marion township of Hocking county, and Jackson and Pike townships of Perry county. These townships are now drained by the Hocking river and its tributary, Rush creek. The river passes diagonally through Berne township, from Lancaster to Sugar Grove. A little below Lancaster it receives a small tributary, Pleasant run, and at Sugar Grove the considerable stream of Rush creek. Pleasant run rises on the drift plain of Pleasant township and flows south into Berne township. In the southern part of Pleasant township its valley is quite deep and of considerable size. On entering Berne township the stream flows out upon a broad, almost level alluvial plain. The stream bears to the westward across this plain for about two miles when it again enters a comparatively narrow valley bounded by high hills, which it follows southward to the Hocking.

Rush creek rises in the vicinity of New Lexington, Pike township, Perry county, at about 871 feet A. T., flows east to Bremen, thence south into Hocking county a short distance and then turns east again into Fairfield county. At Bremen it receives a considerable tributary, North Fork Rush creek, from the north. This branch rises on the drift till plain near Hadley Junction and Pleasantville and flows

southward through a shallow trough cut from the drift until near Rushville when its valley suddenly becomes transformed into a narrow rock
gorge with the hills rising almost vertically 150 to 200 feet from the
stream, with scarcely room for the railroad and the creek between the
rock walls. And as suddenly does it again open out upon the broad
alluvial plain of the Rush creek valley at Bremen. Two other smaller
tributaries of Rush creek must be mentioned. Raccoon creek which
rises near Pleasant run in Pleasant township and flows southward almost parallel to that stream, and only a mile or so east of it, until it
reaches Berne township when, like Pleasant run, it flows out upon a
broad alluvial plain and turns east and flows through a broad and open
valley, which is nearly a mile wide, to Rush creek at Bremen. On the
township line between Pleasant and Berne the rock divide between
Pleasant run and Raccoon creek, is about 150 to 200 feet above the
streams, while in Berne township where Pleasant run turns west and Raccoon creek turns east, the two streams are on the same alluvial plain of a
broad east and west valley. In the early days when the country was
new and this old alluvial plain was timbered the waters of Raccoon
creek joined those of Pleasant run and flowed westward into the Hocking. Their deflection eastward was brought about by the construction
the of an old mill pond and dam. The ditch dug for a waste way from
the mill wheel found a slightly lower level eastward, while the natural
overflow from the mill pond was westward. In time the pond filled with
silt, the mill was abandoned, the wheel (an overshot) decayed, the dam
also rotted away, the pond drained out through the waste way, and
Raccoon creek was added to Rush creek. A few logs still remain in the
bed of Raccoon creek to mark the site of the dam, the banks of the
pond have disappeared under the leveling action of plow and harrow,
but the whole story is told by the logs in the bed of the stream and the
eight feet of silt above the buried soil, which now shows where the
water has cut out its channel through the middle of the old pond.

Turkey creek rises in Monday creek township of Perry county and
flows north-westward to join Rush creek, which is here flowing south
eastward. Its valley is continuous in direction and conforms in depth
and width to that of the Rush creek valley from the point of the con-
fluence of Turkey creek with Rush creek to Bremen.

The topographical features in the vicinity of Lancaster can best be
observed from Mt. Pleasant, a bold bluff of Logan conglomerate just

north of the city. Looking northward and westward the view extends many miles over the broad drift plain of central Ohio. The waters of the Hocking can be seen for many miles. The valleys of the streams are nothing more than shallow troughs cut out of the almost level till plain.

Southward the Hocking rivers enters the hills in a valley about a mile wide. The hills rising 200 feet on each side of the valley. Along this same valley extends the C. H. V. & T. R. R. and the Hocking canal. The valley of the Hocking grows narrower and deep towards the south. Looking eastward there is observed a broad valley equal to if not longer than that of the Hocking and uniting with the latter at Lancaster. For several miles east of Lancaster the valley is not occupied by any stream but is crossed by Pleasant run on its way to the Hocking. This valley is traversed by the almost level track of the C. & M. V. R. R. from Lancaster to Bremen.

A view from one of the hills near Bremen shows that the valley extending from Lancaster to Bremen continues eastward and is occupied by Brush creek. Just north of Bremen the North Fork of Rush creek enters the valley through a very narrow rock gorge. South of the town Rush creek turns south into a valley about three quarters of a mile broad where it joins the larger east and west valley. The observer wonders why the waters of Rush creek should turn into this smaller valley which runs back among the hills and does not continue its eastern course through the broad and open valley to Lancaster. Just east of Bremen the alluvial bottoms are veritable swamps and cover a large area. The old valley seems to have been broadened out here by its lateral tributaries. Passing eastward along the line of the C. & M. V. and T. & O. C. R. R. the valley of Rush creek narrows gradually. Tributary valleys of considerable size enter from both the north and south sides. The Shawnee division of the B. & O. R. R. crosses the valley by following two of these lateral valleys tunneling at both divides. At New Lexington the valley may be said to end. The valley floor is here about 871 feet A. T. The C. & M. V. R. R. turns north up a small branch and about three miles from New Lexington tunnels through the divide. The T. & O. C. R. R. turns south and tunnels the divide within about a mile from the city.

From Bremen southward the valley of Rush creek narrows rapidly and appears continuous with that of Turkey run but the stream follows a small tributary valley and where it crosses the county line into Hocking county it is flowing between almost vertical rock walls 200 yards apart at flood plain. The valley again broadens toward Sugar Grove, the main valley bearing northward while a somewhat lesser valley turns southward both opening at once into the valley of the Hocking.

A few miles above Sugar Grove Rush creek deserts both of these outlets and has cut for .itself a very narrow and picturesque gorge among the hills along the south wall of the old valley. The whole drainage of Rush creek seems to have been determined to run con‌trary to all the laws of hydraulics.

The geological structure of the entire region is upper Waverly and lower Carboniferous sandstones and shales dipping slightly to the south‌east. As far as observed the underlying rock structure has had no in‌fluence in the determination of the drainage lines.

The glacial boundary through this region has been located by Prof. G. F. Wright, at Lancaster. This will serve as a general bound‌dary but local extentions are to be expected. Extending all along the valley east from Lancaster to a half mile beyond Junction City drift de‌posits, consisting of stratified and unstratified gravels and till are scat‌tered at high levels on both north and south walls of the valley. These deposits extend for some distance south of Bremen down Rush creek but none were observed beyond the county line where an old col un‌doubtedly existed.

The most eastern till deposit occurs about ½ mile east of Junction City where it fills the old valley 100 feet above the flood plain of the creek and has caused a slight deflection of the stream to the south around a large island like hill of Waverly. The rail road has here made a deep cut through the till and revealed an excellent section.

Besides the high level deposits which must be attributed directly to the ice. All the large valleys are filled with river wash and silt. At Lancaster this filling in the Hocking valley is at least 220 feet deep. At a point about midway between Lancaster and Bremen rock was reached by a gas well at 175 feet. At Bremen a gas well very much to one side of the valley penetrated 65 feet of filling.

Assuming a uniform grade in the old rock floor from New Lexington to Lancaster the filling at Bremen would reach over 100 feet. At the col on county line on Rush creek, the rock is only about 20 feet below the level of the stream and about 50 feet above the valley floor at Bremen. In Rush creek valley a few miles above Sugar Grove, the rock is at least 118 feet below the surface near the middle of the valley.

From the facts stated it seems certain that the main preglacial drainage of this region was almost directly east from New Lexington to Lancaster. Tributary valleys of considerable size opened into this main valley at various points, notably one from the north at Bremen in the present position of North Fork, Rush creek. Also near same place one from the south which was the westward continuation of the Turkey creek valley.

As the ice tongue which was pushed up this valley as far as Junction City was withdrawn and while yet the mouth of the valley at Lancaster was filled with ice, the waters were ponded back until they rose over a col, located just on the county line between Fairfield and Hocking, where the present Rush creek crosses, and found their way southward into the Hocking. The position of this old col is so plainly indicated by the surrounding topography that its location can be made with certainty to within a hundred yards Just north of this valley the waters ponded back by the main ice front crossed the divide at the col near Rushville and scoured out the gorge from Rushville to Bremen, thus finding an outlet southward along Rush creek.

This torrent of glacial waters had so deepened the Rush creek drainage. line and silted up the old valleys that when finally the ice was withdrawn from the mouth of the valley at Lancaster the waters did not re-occupy their old valleys but followed the glacial drainage lines. The lower portion of the present Rush creek valley, south of the county line, was occupied in preglacial time by a stream heading in several small streamlets in the eastern portion of Marion township, Hocking county. Towards its mouth it bore to the north-westward to join the Hocking and not to the south-westward as at present.

When it is borne in mind that many facts not stated herein seem to indicate that the present Hocking is a reversed stream, it becomes apparent that these preglacial valleys conform to the original preglacial Hocking drainage and add more evidence to the support of the opinion that the preglacial Hocking ran north-westward. And this in its turn to the still larger problem of the central Ohio preglacial river system formed from northward flowing streams which crossed the present course of the Ohio river. One more link is thus added to the chain of evidence in support of the view that the Ohio river along southern Ohio owes its origin and position to glacial forces and does not date back of the glacial period.

Photo and Eng. by W. G. Tight.

The Old Valley. Looking from A, Plate IV, toward Lancaster. Rolling Gravel Hills on the Left.

Photo and Eng. by W. G. Tight.   The Rocks at the Col on the County Line at B, Plate IV.

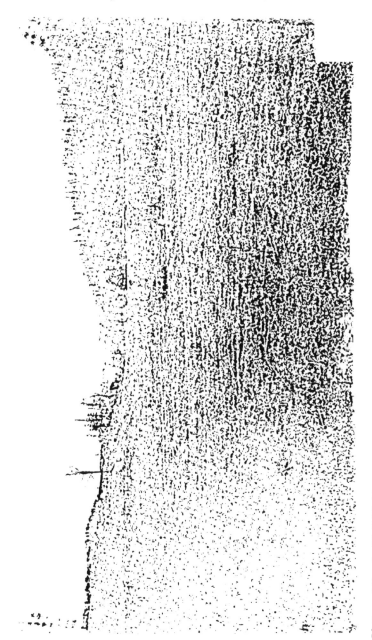

Photo and Eng. by W. G. Tight.
The Col at County Line at C, Plate IV., looking up Rush Creek.   The bluff to the left
marks the top of the old gradation plain.

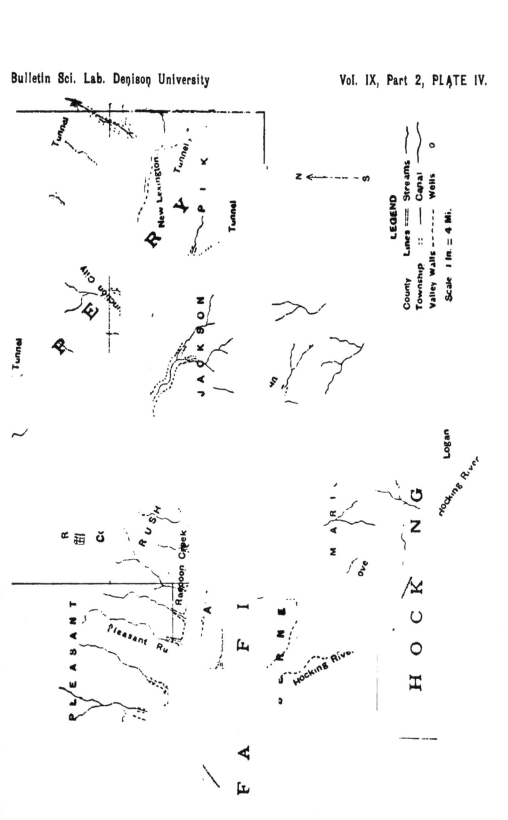

BULLETIN

OF THE

Scientific Laboratories

$11,590$

OF

DENISON UNIVERSITY

VOLUME X.

MEMORIAL VOLUME.

ILLUSTRATED.

———

EDITED BY

W. G. TIGHT, M.S.,

Department of Geology and Natural History.

———

BARNEY MEMORIAL HALL—SOUTH FRONT

BULLETIN

OF THE

Scientific Laboratories

OF

DENISON UNIVERSITY.

VOLUME X.

MEMORIAL VOLUME.
ILLUSTRATED

EDITED BY
W. G. TIGHT, M.S.,
Department of Geology and Natural History.

GRANVILLE, OHIO, AUGUST, 1897.

The University Press.

GRANVILLE, OHIO.

CONTENTS,

LIST OF ILLUSTRATIONS.

BARNEY MEMORIAL SCIENCE HALL—NORTH EAST ENTRANCE

EDITORIAL STATEMENT.

When the Barney Memorial Science Hall was completed and the work fairly begun in the new quarters the advantages presented to the student were so great, as compared to those which the writer had enjoyed in the early days under Professor Hicks, that the fact of the great debt we owe to the laborers of the past and the foundation that they laid was so forcibly presented as to suggest the thought of collecting in some permanent form such facts as could be obtained with reference to the work in science in Denison University and the lives and works of those who wrought so faithfully in the past. In order to show the fruits of their labors it is also important to show the present state of the scientific work and equipment.

In tracing the development of science in this country, and especially the development of the scientific laboratories, the very significant fact is manifest that almost all the large laboratories have been built and equipped through the generosity of broad minded and public spirited men of large fortune.

It is true that the United States government and most of the States have, at public expense, made large investments in the building of experiment stations and research laboratories for scientific work and instruction. Yet it cannot be said that in this country the government is the leading patron of the scientific laboratory. In many other countries the government has taken the leading part in furnishing the means for the pioneer work in the development of the scientific laboratories; but with our form of government, where the mass of the people is the governing power, it is evident that the people must first be shown the benefits to be derived from the establishment of such expensive plants for scientific work before they will vote their money to the support of such enterprises.

Thus it is that while the value of public schools maintained at public expense was early recognized as a necessity and provided for, yet it must be said that the higher institutions of learning have largely been founded and fostered by private means. This is not more marked in any department than that of science.

The founding and maintainance of scientific laboratories and experiment stations as government institutions, at large expense, marks a later stage in the development of scientific work; when the majority of the people have been so educated that they recognize the benefits, to the commonwealth, to be derived from the encouragement of research laboratories then they are willing to vote a tax on themselves for their construction and support.

The development of these public laboratories under governmental patronage has in no way checked the investment of private wealth for the still larger growth of science and scientific instruction. It will probably always be true that the advance steps will be taken in the future as in the past through the patronage of liberal minded men and women of large fortune.

In the growth of the scientific laboratories of this country there are then two factors present - the scientific student, worker, investigator or teacher and the patron of science. While we are remembering the one we can not forget the other.

It is the plan therefore to include in this volume an outline of the development of the scientific work in Denison University, a brief sketch of the lives of those who have been connected with the work, the patron who has so generously placed science here on its present basis, the present condition of scientific instruction and the present instructors, with a short description of the scientific equipment as found in Barney Memorial Science Hall.

The papers thus far published in the Bulletins are largely of a technical character and our excuse (if such is necessary) for introducing this volume in the series is found in the words of Professor C. L. Herrick, the founder of the Bulletin, in his editorial statement to the first volume, where he says, " The Bulletin is intended to represent the life of the college in its scientific departments and may incidentally serve to illustrate to distant friends the facilities for work afforded, as well as needs unsupplied." It is hoped that the matter contained herein will be of interest to our distant friends and that perhaps others may get some new ideas from our equipment as described and illustrated, as we received many from similar volumes furnished us by our distant co-workers.

In the preparation of this volume our obligation is expressed to President D. B. Purinton; Professor W. H. Johnson, of the department of Latin; Professor A. D. Cole, of the department of Physics and Chemistry; Professor C. Judson Herrick, of the department of Biology; and Mr. B. F. McCann, of Dayton, O., for various portions of the historical, biographical, and descriptive text. Credit is given in the text for articles quoted. Also to Professor W. H. Boughton, of the department of Mathematics and Engineering, for the drawings of the floor plans of Memorial Hall; to Mr. L. I. Thayer, a student in photography in the department, for most of the photographs of the interiors of Memorial Hall; to Mrs. Burton Huson for the loan of Professor L. E. Hick's picture from which the cut was made, and to the Board of Trustees for the special appropriation for the expense of publication.

The cuts for the illustrations were all made by the Editor, in the department of Photography and Engraving, and should they not come up to the standard of such work it is hoped the deficiency will be pardoned from the fact that they are the work of an amateur and furnished the Editor's recreation during their preparation.

DENISON UNIVERSITY CAMPUS FROM THE EAST

HISTORICAL SKETCH.

Denison University was founded before the era of the Natural Sciences as an important part of college education began, and therefore not much can be said of scientific studies at Denison in its earlier history. The primary object in the minds of its founders had been to make provision for educating Ministers of the Gospel, in order to facilitate the evangelization of Ohio's rapidly increasing population. It was recognized, however, that the young minister needed something more than a purely theological training; and what the predominating character of that additional study was to be was indicated in the name chosen for the school, — *The Granville Literary and Theological Institution*.

For more than twenty years from the beginning the Faculty rolls show no closer approach to a recognition of the Natural Sciences than the title Professor of Mathematics and *Natural Philosophy*. An examination of the catalogues, however, shows that one term's work was generally given to Chemistry and one to Geology and Mineralogy. It is evident that these branches were taught as extras by the Professors in other departments, and under such circumstances the instruction must have been confined largely to text-book work.

In 1853, a Professorship of Natural Sciences was instituted, but disappeared after a single year, the incumbent, Professor Fletcher O. Marsh, being transferred to the chair of Mathematics and Natural Philosophy. During this year, however, Botany, Anatomy and Physiology were added to the curriculum, and a separate Scientific Course was published, with the promise that the degree of Bachelor of Science would be conferred upon those by whom it should be taken. It was stated that this course was "designed to furnish a suitable education to those who are fitting themselves for business men, for engineers, or to engage in mercantile or mechanical pursuits." This course was one year shorter than the Classical, and contained no work in any language other than English.

In the catalogue of 1854 appears the first announcement of an Agricultural Department, which deserves mention here as indicating good intentions in the direction of scientific work, although such intentions were never carried into effect. The announcement was as follows: "The Trustees have resolved to establish an Agricultural Department, in which the best facilities for obtaining a knowledge of the science of agriculture shall be afforded, consisting of lectures, experiments and general instruction in those sciences which have a more direct relation to this important branch of industry; embracing a period of fifteen weeks, during the second term of each year. Instruction in this department will * * * afford the sons of farmers, and others, an opportunity to spend the winter months in listening to lectures from scientific professors and practical agriculturists, who will be employed to give instruction in Agricultural Chemistry, embracing the nature of the soil in this state, and its adaptation to the different productions of this latitude, and the best means by which the different kinds of soil may be enriched; in Practical Mechanics, embracing the structure of buildings, fences and farming tools, with referenence to durability, utility and economy; In Geology, embracing the mineral resources of the State; in Agricultural Jurisprudence, embracing the laws relating to deeds of conveyance, trade and agricultural pursuits; in Animal and Vegetable Physiology, embracing the kind of animals adapted to the climate, the best methods of rearing them, the diseases to which they are subject, their comparative expense and the means of their improvement, and the culture of the different kinds of grain and fruit."

From the period just considered to the year 1870 no new scientific studies were added to the curriculum, and during a portion of the time Botany disappeared. In the enlargement of the Faculty which followed the completion of a new endowment fund, in 1867, no Professor or Instructor in science was added, the small amount of scientific work provided for in the curriculum still remaining in the hands of the occupants of other chairs. In the catalogue of 1870, however, appears the name of Lewis E. Hicks as Professor of the Natural Sciences and it is only from this date that Denison can fairly be said to have comprised a Scientific Department of study. This addition affected immediately the Classical as well as the Scientific Course. Besides the Chemistry, Geology and Mineralogy, and Anatomy, Physiology and Hygiene of former years, Classical students were now required to take " Natural History " and Vegetable Physiology in the Sophomore year, and Zoology in the Junior. Vegetable Physiology and Zoology are the only additions which immediately appeared in the Scientific course, though the results accomplished in the branches before taught were doubtless more satisfactory now that these branches were in the hands of one who could give his time almost entirely to scientific work. We say *almost*, since Professor Hicks was compelled by the exigencies of the situation to do some work in unrelated lines of study, just as others had been compelled, previous to his appointment, to do scientific work. German and French were first added to the Scientific Course during this year. It was not until 1881, however, that the course was lengthened from the three year limit and made equal to the Classical Course in the number of years of collegiate study' required. This equality of time, however, was more apparent than real until 1886, since the requirements for entrance to the Freshman class were less exacting by one year's work than for Classical students until that date. This shortness of the Scientific Course did an injury to the development of scientific work at Denison even greater than the deficiency of time, in that it furnished a refuge for students who fell behind in their Classical work. As long as this condition continued the presence of a small element of such men in the Scientific Department undoubtedly tended to deter bright students from becoming candidates for the Bachelor of Science degree. Of course there were those whose preference for Scientific work was sufficently strong to cause them to disregard this feeling, but the experience of the last ten years has shown beyond a doubt that the course has become much more popular by being made longer and harder. It is a significant fact that the phrase "*gone Scientific,*" is no longer understood in student parlance as an equivalent for "failed in Greek and Latin."

It is due to Professor Hicks, of course, to say that this condition of affairs was decidedly contrary to his own desires in the matter. He would gladly have lengthened and strenthened the work in Science if the income of the University had been sufficient to provide the necessary additional teaching force and equipment. A great deal of illustrative material was accumulated by his personal efforts, and by others under his direction, which could not be used to advantage during his term of service because of the narrow quarters in which the work of the department had to be carried on. Under present conditions this material is now largely available for the practical purposes of instruction, and thus an important portion of his labor for the University is now bearing its first fruits.

PROFESSOR LEWIS EZRA HICKS, A.M.

Professor Hicks was born at Kalida, Putnam County, Ohio, March 10, 1839. He had not yet completed his college education when the War of the Rebellion began, but he fought in the Union army during the whole four years, serving as Lieutenant Colonel in the 69th O. V. I. After the close of the war, he completed his college course at Denison, doing some teaching as an Assistant in the Preparatory Department at the same time and graduating with the A. B. degree in the class of 1868. During the following year he remained as a Tutor in the Classics. He then went to Harvard for a year to pursue special work in Zoology and Geology, where he had the good fortune to be a student under Louis Agassiz.

In 1870, he came back to Denison as Professor of Natural Sciences, and remained until 1884, when he resigned to accept the chair of Geology in the University of Nebraska. During the last two years of his service at Denison, the title of his Professorship was changed to Geology and Natural History, in view of the endowment of a chair of Chemistry and Physics, by the Chisholms, of Cleveland. Professor Hicks retained his chair in the University of Nebraska until 1891, and during a portion of this time was also connected with the United States Department of Agriculture, as Assistant Geologist. He was a member and fellow of the American Association for the Advancement of Science; a member of the American Society of Irrigation Engineers; one of the founders of the Geological Society of America, and a Fellow of the same, as well as one of the founders of the *American Geologist* and long an Associate Editor. From the 1893 edition of the General Catalogue of Denison University we take the following list of his contributions to scientific literature, a list not intended to be exhaustive : " Scientists and Theologians : How they Disagree, and Why," a series of articles in the Baptist Quarterly Review, 1874; " A Critique of Design Arguments," an octavo volume of 417 pages published by the Scribners in 1883; " Discovery of the Cleveland Shale in Central

PROFESSOR LEWIS EZRA HICKS, A. M.

Ohio," American Journal of Science, 3d Series, Vol. 16, p. 70; "The Waverly Group in Central Ohio," ib., p. 216; "The Dakota Group in Nebraska," Proceedings of the American Association for the Advancement of Science, Vol. 34; "Irrigation in Nebraska," Bulletin No. 1, of the Agricultural Experiment Station of Nebraska; an article on the same subject in the Report of the Nebraska State Board of Agriculture, 1887, p. 122; "The Soils of Nebraska," (with a geological map of the state), Report of the State Horticultural Society of Nebraska, 1888; "The Permian in Nebraska," Proceedings of the American Association, Vol. 36, p 216; an article on the same subject in the American Naturalist, Vol. 20, p. 881; "Geology in its Relations to Agriculture," Report of the State Board of Agriculture, 1889, p. 364; "Silting, or Basin Irrigation," ib., 1890, p. 131; "Storage of Storm Waters on the Great Plains," ib., 1891, p. 172; "An Old Lake Bottom," Bulletin of the Geological Society of America; "Artesian Wells in Nebraska," Senate Executive Document, 222, 51st Congress (with geological map of Nebraska); "Soils and Waters of the Lake Region, as Related to its Geological Structure," Report of the Nebraska Board of Agriculture, 1892; "Irrigation and Horticulture," Report of the State Horticultural Society, 1892, p. 78; "Tree planting in Canons," ib , 1893; "Evolution of the Loup Rivers," Science, Vol. 19, No. 469, Jan. 29 1892; "Some Elements of Land Sculpture " Bulletin of the Geological Society of America, Vol. 4, p 133; "Irrigation in Nebraska," Senate Executive Document, 41, Part III, 52nd Congress, First Session.

In addition to his scientific work, Professor Hicks maintained always a lively sense of his responsibilities as a member of society and as a citizen. He took a deep interest in all political questions and was entirely independent of party dictation at a time when independence was not yet common, a fact which made it inevitable that his political position should sometimes be misunderstood. In Lincoln, Nebraska, his activity in municipal politics resulted in his elevation to the Chair manship of the Board of Public Works. He is now engaged in college work in Burmah.

PROFESSOR ISAAC JUSTUS OSBUN, A.B.

The work of the above mentioned Chair of Chemistry and Physics (founded in memory of Henry Chisholm, of Cleveland, by his children) was divided up and assigned to other Professors for a year and then placed in charge of Isaac J. Osbun, as Professor of Chemistry and Physics.

Professor Osbun was born in Windsor, Ohio, May 19, 1850. He was for six years a student in Granville, entering in the Preparatory Department in 1866 and graduating in the Classical Course, with the class of 1872. After a year's work as Principal of the Glendale High School he went to Europe and spent two years as a student in the Universities of Stuttgart, Tuebingen, Heidelberg and Paris. Upon his return he was chosen Principal of the Berkshire Institute, New Marlborough, Massachusetts, but gave up this position a year later to accept the Professorship of Chemistry and Physics in the State Normal School at Salem, Massachusetts, where he remained for seven years, resigning to take his Professorship at Denison, in 1883. Here he died, December 8, 1884, in the first term of the second year of his work. We include here a number of extracts from an article written for the Denison Collegian soon after his death by Dr. W. C. Davies, between whom and Professor Osbun there had existed a very intimate friendship from his student days until the end of his life:

" During his college course, he displayed great love for the sciences. Not content to blindly accept the statement of the text-book or teacher, he wanted to work out principles for himself. Lack of apparatus he did not allow to become a hindrance, but transformed his room into a workshop. The writer of this article well remembers many a piece of home made apparatus which he borrowed to demonstrate the principles of Physics to his own pupils. [Dr. Davies was then in charge of the Granville schools.] Though home-made, they always answered the purpose for which they were made, and gave evidence of the originality and skill which, in later years, found a wider

PROFESSOR ISAAC JUSTUS OSBUN, A. B.

field of operation. The child is father to the man, and these traits which marked his life as a student became important characteristics of his work in teaching. The enforced dependence upon himself for means to demonstrate what he would not accept without demonstration, was valuable training for his future work." [The compiler of this article well remembers, as a student under Professor Osbun, the habit of insisting upon actual demonstration to which Dr. Davies here calls attention. It often seemed irksome to be required to perform a series of experiments in the laboratory in order to demonstrate to the eye some principle which presented no difficulty whatever to the mind, and had perhaps been understood and accepted as almost self-evident long before; but it is easy to see now that this was done not primarily for the sake of the particular point involved in the experiment but to establ sh the experimental habit more firmly in the character of his pupils.]

" During his college course, Mr. Osbun was a faithful student, leading his class in all studies that were congenial to him. He was an earnest and active member of his literary society, especially liking, and excelling in, debate * * * Of Professor Osbun's life and work since he returned to us, a little more than a year ago, much *might* but little *need* be said. They speak for themselves. Not only do his associate teachers and the students mourn his loss, but his death touched a chord which vibrated through the whole community. Measured by years, his life was short. At thirty-four, we look upon a man's life as only fairly begun. Measured by what he accomplished, a man of three score would have no cause to blush. His life was one of unceasing activity. To be idle was to him simply impossible. During the vacation following his Junior year, in a country school house a few miles from Granville, he delivered his first scientific lecture, illustrated by experiments of his own devising. It was the first of a long list. When teaching at Salem he delivered as many as sixty lectures in one year. These lectures embraced a great variety of subjects, and, while some were delivered before popular audiences, many were before the most cultivated and critical scientific associations. He delighted to choose some simple subject and lecture on it before those destitute of scientific training ; and the ability he displayed of clothing the bare facts of science with so much interest that he fascinated as well as instructed even the most ignorant of his hearers, was signal proof of the originality of his mind and thoroughness of his preparation. Even va-

cation was not a time of rest to him.   One summer he spent in aiding
Professor Bell to perfect the telephone.   During another, that of 1882,
he lectured at Martha's Vineyard, before the Summer School of Sciences   In fact he was in almost constant demand at Teachers' Insti-
tutes and Associations.   He wrote many articles on scientific subjects,
for publication.   *   *   *

" His heart was in his work to the very close of life.   The dying
warrior on St. Helena in his delirium imagined himself at the head of
his army, and our teacher carried on his work to the last.   The first
indication that his mind was " wandering " came at midnight, in a
direction, clear and sharp, to his class, in regard to the performance of
an experiment.   His mind, released from the control of the will, was
true to itself and its chosen work to the very end.   As a teacher he
attained the very highest success, and this success which crowns his
life was the legitimate reward of straightforward, earnest, well directed
and persevering toil.   He magnified his work, and the results of his
work, written in the minds and hearts of those whom he taught, constitute a monument to him more enduring than granite."

Epoch making work is not always at once realized as such, but it was
easily seen at the time that Professor Osbun's year at Granville had inaugurated a new epoch in the scientific work of Denison University
An editorial in the Collegian said :

" Professor Osbun, feeling that his work could not be a success
without certain fundamental improvements, made his coming condi-
tional upon their supply.   He has been on the field a little more than
a year, and a revolution has already been wrought whose effects will
be seen in all coming years.   Never again, under any circumstances,
can the work of this department be what it was before.   All his training and experience had taught him what must be done that the work
might be worthy of the college and himself, and with the persistency
which was his characteristic, he toiled to gather and utilize everything
that could be reached."

The curriculum as it stood during the year of Professor Osbun's
death contained the following scientific work : candidates for the degree of Bachelor of Science were required to take Chemistry during
the whole of the Freshman year and one term of the Junior year
Physiology was required during the Winter term of the Sophomore year,
and Botany, with some other scientific study to be chosen, during the
Spring.   The work in Botany was continued during eight weeks of the

Fall term of the Junior year and this was followed by Zoology during the remainder of the term. In the Senior year, Geology was taken during the Fall term and Paleontology during the Winter. About half as much scientific work was required of students in the Classical and Philosophical Courses. With the addition of new studies and appliances, the amount of floor space devoted to the work had been increased more than threefold by the sacrifice of dormitory rooms on the first and second floor of the " New Brick," and thus the work of instruction was made much more effective than it had been in former years A temporary water supply for the work of the laboratories had been secured by drilling a well on top of College Hill and putting in a wind pump, but this never furnished a satisfactory supply and was soon superseded by the Granville water works, constructed during the year 1885-6. For the remainder of the year of Professor Osbun's death, the work of the department of Chemistry and Physics was done by Nathan F. Merril, Ph D. At the opening of the following year, Professor A. D. Cole, a graduate of Brown, who had been doing Post-Graduate work at Johns Hopkins, took charge of the department and still remains, having been absent one year pursuing special researches in Physics at the University of Berlin. During his absence, the work of the chair was in the hands of Mr. E. P. Childs, a graduate of Denison, and now Professor of Science in the High School of Pueblo, Colorado. The development of the Department under Professor Cole made necessary in 1896 the employment of an Assistant, Mr. H. C. McNeil, who graduated from Denison in the Scientific Course with the class of that year. Previous to the endowment of this Department, some work in Physics had been done by Professor Gilpatrick, in addition to his work in Mathematics, and still earlier by Professor Marsh. The work in Astronomy, done by the Professor of Mathematics, still connects this chair closely with the scientific portion of the Faculty; and the Assistant in Mathematics, Mr. W. H. Boughton, gives a portion of his time to instruction in Physics and Chemistry

PROFESSOR CLARENCE L. HERRICK, M.S.

To return to the chair of Geology and Natural History, after the resignation of Professor Hicks the University had the good fortune to secure the services of Professor Clarence L. Herrick, who had been employed for the work temporarily during a short visit of Professor Hicks to Europe for some special researches in the British Museum. Professor Herrick remained at Denison until 1889, then accepted the Professorship of Biology and Geology in the University of Cincinnati, which he held for three years, and then came back to Denison as Professor of Biology, remaining in active charge of ,the chair until 1894, when ill health compelled him to seek the climate of New Mexico. At the close of the last school year, very much to the regret of all concerned, he resigned his chair, in view of continued inability to endure the climate of Ohio with safety. Since that date, he has been chosen as President of the University of New Mexico and has entered upon his work.

Professor Herrick was born in Minneapolis, in 1858, and graduated from the University of Minnesota in 1880, where he remained as Instructor in Botany and Zoology during the following year. He spent the year 1881-2 in study in Europe, and then accepted the position of State Mammologist, in connection with the Geological Survey of Minnesota, in which work he remained until called to Denison.

Beginning upon the foundations which we have described, Professor Herrick gave an enormous impetus to all branches of scientific study. In spite of any adequate financial provision for such work, he began immediately the publication of the " Bulletin of the Laboratories of Denison University," which has now reached its tenth volume and has been of inestimable value to the University in stimulating original research, by furnishing an avenue for the publication of results, in calling the attention of scientists all over the educational world to the character of work done here, and in bringing to the Library by exchange a mass of scientific literature which could have been secured in no other

PROFESSOR CLARENCE L. HERRICK, M. S.

way. While in the University of Cincinnati, he founded the *Journal of Comparative Neurology*, which was adopted as one of the official publications of Denison upon his return here in 1892, and has maintained a very high standing among neurologists ever since its appearance. It is now a joint publication of Denison and Columbia Universities, Dr. Strong of the latter being one of the editors. Its work, however, together with that of the Bulletin, is more fully described elsewhere in this volume.

Among Professor Herrick's literary and scientific contributions, exclusive of articles in the BULLETIN OF THE SCIENTIFIC LABORATORIES and the *Journal of Comparative Neurology*, the following partial list includes his principal writings:

Microscopic Entomostraca. *Geol. and Nat. Hist. Surv. Minn.* Seventh Ann. Rep., 1879.
Fresh-water Entomostraca. *American Naturalist*, 1879.
List of Birds of Minnesota.
List of Mammals from Big Stone Lake, with new sp., etc., *Annual Rep. Geological Survey of Minnesota.*
Types of Animal Life. A Laboratory Hand-Book. *Minneapolis, Minnesota,* 1881.
Habits of Fresh Water Crustacea. *American Naturalist*, 1882.
A New Genus and Species of the Crustacean Family of Lyncodaphnidæ. *American Naturalist*, 1882.
.Papers on the Crustacea of the Fresh waters of Minnesota. I. Cyclopidæ of Minnesota. II. Notes on some Minnesota Cladocera. III. On Notadromas and Cambaras. *Tenth Ann. Rep. Geol. Surv., Minn.*, 1882.
Heterogenetic Development in Diaptomus. *American Naturalist*, 1883.
Hetrogenesis in the Copepod Crustacea. *Am. Naturalist*, Feb., 1883.
A new Genus and Species of the Crustacean Family Lyncodaphnidæ. *Am. Naturalist*, Feb., 1883.
A blind Copepod of the Family Harpacticidæ. *Am. Naturalist*, Feb., 1883.
A Final Report on the Crustacea of Minnesota included in the orders Cladocera and Copepoda. *Geol. and Nat. Hist. Surv. Minn.* 1884.
Outlines of Psychology : Dictations from Lectures by Hermann Lotze. Translated with the addition of a Chapter on the Anatomy of the Brain. *Minneapolis, Minn.*, 1885.
Contribution to the Fauna of the Gulf of Mexico and the South. *Mem. Denison Scientific Assoc.* Vol. I, No. 1. 1887.
Some American Norytes and Gabbros. *American Geologist*, Vol. I, No. 6, 1888. With E. S. CLARK and J. L. DEMING.
Science in Eutopia. *American Naturalist*, 1889.
Notes upon the Waverly Group in Ohio. *American Geologist*, Vol. III.

A Contribution to the Histology of the Cerebrum. *The Cincinnati Lancet-Clinic*, Sept. 28, 1889.

Modern Thought and Modern Faith   *The Standard*, 1890.

Additions and Corrections to Miller's North American Palæontology. *American Geologist*, Vol. V, No. 4, 1890.

Notes upon the Brain of the Alligator. *The Journal of the Cincinnati Society of Natural History*, Vol. XII, 1890.

Suggestions upon the Significance of the Cells of the Cerebral Cortex. *The Microscope*, Vol. X, No. 2, 1890.

The Commissures and Histology of the Teleost Brain. *Anatomischer Anzeiger*, 1891.

Additional Notes on the Teleost Brain. *Anatomischer Anzeiger*, 1892.

Notes upon the Histology of the Central Nervous System of Vertebrates. *Festschrift zum siebenzigsten Geburtstage Rudolf Leuckart's*, 1892.

Mammals of Minnesota. *Bull. VII, Geol. Surv., Minnesota*, 1892.

Notes upon the Anatomy and Histology of the Prosencephalon of Teleosts. *American Naturalist*, Feb., 1892.

Wood's Reference Hand-Book of the Medical Sciences, Vol. IX, Suppl. 1893. Articles as follows :

     1.   The Comparative Anatomy of the Nervous System.

     2.   The Histogenesis of the Elements of the Nervous System.

     3.   The Physiological and Psychological Basis of the Emotions.

     4.   Waller's Law.

The Scope and Methods of Comparative Psychology. *Denison Quarterly*, Vol. I, Nos. 1-4, 1893.

Synopsis of the Entomostraca of Minnesota. *Second Report of the State Zoologist of Minnesota*, 1895. With C. H. TURNER.

Microcrustacea from New Mexico. *Zoologisher Anzeiger*, No. 467, 1895.

The Testimony of Heart Disease to the Sensory Facies of the Emotions. *Psychological Review*, III, 3, 1896.

Suspension of the Spatial Consciousness. *Psychological Review*. III, 2, 1896.

The Critics of Ethical Monism. *The Denison Quarterly*, Vol. IV, No. 4, 1896.

The Psycho-sensory Climacteric. *Psychological Review*, III, 2, 1896.

The So-called Socorro Tripoli. *American Geologist*, 1896.

The Geology of a Typical Mining Camp. *American Geologist*, XIX, 4, 1897.

The Propagation of Memories. *Psychological Review*, IV, 3, 1897.

The Waverly Group of Ohio. *Final Report Geol. Surv. Ohio*. Vol. VIII.

The Waverly Group of Ohio, *Bulletin of the American Geological Association*.

Inquiries regarding Current Tendencies in Neurological Nomenclature. *Journal of Comparative Neurology*, Vol. VII, No. 3, 1897. With C. JUDSON HERRICK.

Articles in *Baldwin's Dictionary of Philosophy and Psychology*, The Macmillan Co. [In Press.] With C. JUDSON HERRICK.

The growth of the work in Professor Herrick's department, both in the College and Academy, made it necessary to secure assistance in 1888, and W. G. Tight, a graduate of Denison in the class of 1886, was employed as Instructor in the Academy. During Professor Herrick's connection with the University of Cincinnati and studies abroad, Professor Tight had charge of the collegiate work. Upon Professor Herrick's return to Denison the work of the department of Natural History was divided, Professor Herrick being Professor of Biology, and Professor Tight being Assistant Professor of Geology and Natural History. Later Professor Tight was given full charge of his department under the title of Professor of Geology and Botany. Since 1890, Professor Tight has been Editor of the BULLETIN OF THE SCIENTIFIC LABORATORIES, mentioned above as founded by Professor C. L. Herrick.

Upon Professor C. L. Herrick's resignation in 1897, the department was placed in charge of his brother, Professor C. Judson Herrick, who had performed a large part of the duties of the chair since January, 1894, with the exception of a year spent in special work at Columbia University, during which time the work was conducted by Mr. H. H. Bawden.

PRESENT CURRICULUM.

It is scarcely necessary to trace the growth of the curriculum step by step during the past ten years. It will serve all purposes to show its present condition, as found in the current catalogue of the University. Applicants for admission to the Freshman class in the A. B. course must have to their credit one term's work in Physical Geography, one in Physiology and one in Elementary Physics; for the Philosophical course, the same, plus an additional term in Physics and one in Botany; for the Scientific course, there is added still further one term's work in Chemistry, one in Anatomy and Physiology, and one in Mechanical Drawing. For the college work, we have thought it well to give the Classical and Scientific courses substantially complete, in order to show the amount of scientific work in each, in relation to other studies. Scientific studies appear in bold-baced type, in order to facilitate the work of comparison. No doubt the line between what is classed as Scientific and what is not will seem arbitrary in some cases, but that is hardly to be avoided.

For the Degree of Bachelor of Arts.

FRESHMAN CLASS.

FALL TERM.

Latin.—Cicero, De Senectute or De Amicitia, followed by Livy, Books I-II, or XXI-XXII ; The Latin Subjunctive.

Greek.—Select Orations of Lysias ; History of Athens under the Thirty Tyrants and Restoration of the Democracy.

Mathematics —Part III of Olney's University Algebra.

Rhetoric.—Hart's, with Lectures, one hour a week.

---

* Separate entries in the courses are in many cases here abbreviated, but in no case omitted.

WINTER TERM.

Latin.—Livy or Sallust, followed by Cicero, De Officiis, or Select Letters.
Greek.—Herodotus and Thucydides.
Mathematics.—Part III of Olney's Geometry ; Plane Trigonometry.
American Literature.—One hour a week.
Rhetoric.—Hart's, with Lectures.

SPRING TERM.

Latin.—Horace, Odes and Epodes.
Greek.—Homer's Iliad; Peculiarities of the Epic Dialect.
Chemistry.—Experimental Lectures ; Recitations ; Laboratory Study of the Non-metals.
American Literature.—One hour a week.
Rhetoric.—Hart's, with Lectures.

SOPHOMORE CLASS.

FALL TERM.

Greek.—Demosthenes ; Greek New Testament, one hour a week.
Mathematics.—Olney's Trigonometry, Plane and Spherical ; General Geometry begun.
Rhetoric.

WINTER TERM.

Latin.—Tacitus, Germania and Agricola, or the Annals, or the Histories ; Letters of Pliny.
Mathematics.—General Geometry and Differential Calculus ; Lectures on the Integral Calculus, four hours a week.
Physiology.—Martin, four hours a week.
English Literature.—Lectures ; Select Reading, two hours a week.
Rhetoric.—Orations.

SPRING TERM.

Greek.—Apology and Crito of Plato ; Greek New Testament, one hour a week.
Botany.—Gray's Manual ; Elements of Plant Physiology.
French.—Whitney's Brief Grammar ; Introductory Reader.
Rhetoric.—Essays and Orations.

JUNIOR CLASS.

FALL TERM.

Latin.—Rhetoric and Literary Criticism among the Romans ; The Dialogus of Tacitus, Book X of Quintilian and the " Literary Epistles" of Horace.
German.—Joynes-Meissner's German Grammar (Lessons I-XXXVI); Brandt's German Reader.
Rhetoric.—Essays.

Mechanics. —Carhart's University Physics, Vol. I.
French.—Super's Historical Readings, last half of the term.
Spanish.—Manning's Spanish Grammar ; Knapp's Spanish Readings.

WINTER TERM.

A Science.*—(Zoology, Chemistry, or Physics.)
Logic.—Davis' Inductive and Deductive Logic ; Method ; Notes on the History of Logic ; Fallacies.
Rhetoric.—Essays, Studies in Shakespeare.

ELECTIVES.

Greek —Tragedies of Aeschylus and Sophocles.
German.—Joynes-Meissner's Grammar ; Schiller's *Wilhelm Tell ;* Müller's *Leitfaden zur Geschichte des deutschen Volkes;* Harris' German Composition ; Dictation and Sight Reading.
Spanish.—Knapp's Readings ; Dictation and Sight Readings ; Selections from Galdos and Valera, Lope de Vega and Calderon ; Berlitz's Exercises.
Mathematics.

SPRING TERM.

A Science.—(Zoology, Cryptogamic Botany, Chemistry, or Physics.)
A Language —Either
   Latin.—The Roman Stage ; Plautus and Terence.
   French.—Erckmann-Chatrian's *Le Conscrit de 1813 ;* Halévy's *L'Abbé Constantin,* and Duval's *Histoire de la Littèrature,* or
   German.—Müller's *Leitfaden zur Geschichte des deutchen Volkes ;* Riehl's *Burg Neideck ;* Harris' German Composition.
Rhetoric.—Orations.

ELECTIVES.

Astronomy.—Young's General Astronomy ; Lectures.
History.—Emerton's *Mediaeval Europe.*
English:—English Literature in the Nineteenth Century.

SENIOR CLASS.

FALL TERM.

Psychology.—Lectures.
History of Philosophy. —Weekly Lectures throughout the Fall and Winter Terms.
English Literature.—Lectures and Select Readings.

---

* The science elected this term for the first time must be continued through the Spring Term, except that Cryptogamic Botany may be substituted for the second term in Zoology.

Geology.—LeConte ; Laboratory and Field Work.
American Politics.—Johnston's *History of American Politics.*
French.
German.
Spanish.

WINTER TERM.

Ethics.—Lectures on Theoretical and Practical Ethics ; Notes on the Philosophy of Ethics and the Moral Code.
Economics.
Rhetoric.—Orations.

ELECTIVES.

German.—Lessing's *Minna von Barnhelm;* Goethe's *Hermann und Dorothea;* Freytag's *Die Journalisten* ; Collar-Eysenbach's German Lessons; Composition and Dictation ; Themes on German History.
French.—Thier's *Bonaparte en Egypte;* Vacquerie's *Jean Baudry;* Masson's *Lyre Francaise;* Chassang's Grammar ; Outlines of History, 1789-1848.
Italian.—Grandgent's Italian Grammar; Harper's *Principia Italiana,* Part II, or Bowen's or Montague's Reader.

Physiological Psychology.

SPRING TERM.

Evidences of Christianity.—Purinton's Theism.

ELECTIVES.

History of Civilization.—Guizot.
International Law.
Italian.—Grandgent's Italian Composition ; Readings from Dante, Manzoni, Pellico.
French.
German.
English.

II.

For the Degree of Bachelor of Science.

The courses of study leading to the degree of Bachelor of Science are based upon the same schedule and are similar in extent, but differ in the amount of time given the characteristic or leading subject.

These courses are : First, a course in Biology [B]; second, a course in Chemistry [C]; third, a course in Geology [G] ; and fourth, a course in Physics [P].*

---

* Bracketed initials signify that subjects so marked are required in the course thus indicated.

Differentiation begins with the Junior Year. Students entering for the Degree of Bachelor of Science must select the course to be pursued before that time and will be permitted to deviate from it only by Faculty vote.

FRESHMAN CLASS.

FALL TERM.

Chemistry.—Qualitative Analysis; Laboratory Courses with weekly recitations on the Chemistry of Metals.
†*French.*—Chassang's Grammar; Duval's *Histoire de la Littérature;* Corneille's *Horace;* or DeVigny's *Cinq Mars;* Herdler's Scientific French Reader; Outlines of French History to 1789.
Mathematics.—University Algebra, Part III.—Olney.
Rhetoric.—Hart's, with Lectures.

WINTER TERM.

Chemistry.—Qualitative Analysis finished; Organic Chemistry, three times a week.
Mathematics.—Geometry; Plane Trigonometry.
French.—Thier's *Bonaparte en Egypte;* Vacquerie's *Jean Baudry;* Masson's *Lyre Francaise;* Branson's *Everyday French;* French History from 1789 to 1848.
American Literature.—One hour each week.
Rhetoric.—Hart's, with Lectures.

SPRING TERM.

French.—Sandeau's *Mademoiselle de la Seiglière;* Effinger's *Sainte Beuve;* Dumas' *Les Trois Mousquetaires;* Branson's *Everyday French;* Gaston Paris' *Chanson de Roland;* French History since 1848.
Chemistry.—Qualitative Analysis.
Dynamic Geology.
American Literature.—One hour each week.
Rhetoric.—Hart's, with Lectures.

SOPHOMORE CLASS.

German.—Joynes-Meissner's German Grammar, (Lessons I–XXV); Brandt's German Reader; Dictation and Composition.
Mathematics.—Plane and Spherical Trigonometry; Olney's General Geometry begun.
Rhetoric.

† Second Year Course prescribed for Freshmen who elect French in the Senior Preparatory year.

WINTER TERM.

German.—Joynes-Meissner's Grammar; Schiller's *Wilhelm Tell;* Müller *Leitfaden zur Geschichte des deutschen Volkes;* Harris' German Composition; Dictation and Sight Reading.

Mathematics.—General Geometry and Differential Calculus ; Lectures on the Integral Calculus, four hours a week.

Zoology.—Vertebrates, Lectures and Laboratory work, four hours a week.

English Literature.— Two hours a week.

Rhetoric.

SPRING TERM.

German. - Müller's *Leitfaden zur Geschichte des deutschen Volkes;* Riehl's *Burg Neideck;* Dictation and Sight Reading.

Zoology.—Invertebrates, Lectures and Laboratory Work.

Mathematics —Surveying.

Rhetoric.—Essays and Orations

JUNIOR CLASS.

FALL TERM.

Comparative Anatomy and Histology.—[B].
Mineralogy.—[C—G].
Calculus. (Half Term); Physical Laboratory.—(Half Term). [P.]
Mechanics.—Recitations and Laboratory work; Lectures on Sound
Rhetoric.—Essays.

ELECTIVES.

German.—Freytag's *Doctor Luther;* Schiller's *Das Lied von der Glocke* and *Ballads;* Collar-Eysenbach's German Lessons throughout the year ; Spanhoofd's *Deutsche Grammatik.*

Spanish.—Manning's Spanish Grammar ; Knapp's Spanish Readings.

Botany.—Structural.

A Science.—Other than that prescribed.

WINTER TERM.

Neurology.—[B.]
Organic Chemistry.—Laboratory work. [C].
Physics—Magnetism and Electricity; Lectures and Recitations, Laboratory Work three times a Week. [P].
Logic. Davis' Inductive and Deductive Logic.
Geology.—Physiographic. [G].
Rhetoric. Essays ; Studies in Shakespeare,

German.—Lessing's *Minna von Barnhelm;* or Schiller's *Die Piccolomini ;* Dippold's *A Scientific German Reader;* Composition and Dictation ; Themes on German History.

Mathematics. – Bridge Construction.   •

Spanish.—Selections from Galdos and Valera; Lope de Vega and Calderon; Knapp's Readings, and Berlitz's Exercises.

A Science.—(Other than prescribed).

Laboratory Physics. – Electrical Measurements. [P].

Botany.—Bacteriology.

SPRING TERM.

Botany.—Cryptogams. [B-G].

Embryology.—Lectures and Laboratory Work. [B].

Chemistry—Laboratory; Advanced Quantitative. [C].

Physics.—Heat and Light; Lectures, Recitations, Laboratory Work. [P].

Astronomy.—Young's General Astronomy; Lectures. [Elective for B].

Rhetoric.—Orations.

History.—Emerton's *Mediaeval Europe.*

German.—Goethe's *Hermann und Dorothea;* Von Sybel's *Die Erhebung Europas gegen Napoleon I ;* Dictation and Composition.

A Science.—(Other than required).

English.—English Literature of the Nineteenth Century.

Botany.—Physiological.

Mathematics.—Strength and Resistance of Materials.

SENIOR CLASS.

FALL TERM.

Psychology.—Lectures.

English Literature.—Lectures and Select Readings.

History of Philosophy.—Weekly Lectures.

Geology.—[G].

Chemistry, Assaying, or Water Analysis.

Physics. – Electrical Engineering. [P].

An Elective [B].

WINTER TERM.

Ethics.—Lectures on Theoretical and Practical Ethics ; Notes on the Philosophy of Ethics and the Moral Code.

History of Philosophy.—Weekly.

Physiological Psychology.—[B].
Technological Chemistry.—[C].
Geology.—Lithology or Paleontology. [G].
Laboratory Physics.—[P].
Rhetoric.—Orations.

ELECTIVES.

Economics.

German.—Goethe's Prose ; History and Literature ; Dictation and Composition.

Italian.—Grandgent's Italian Grammar ; Harper's *Principia Italiana,* Part I., or Bowen's or Montague's Reader.

SPRING TERM.

Evidences of Christianity.—Purinton's Theism.

ELECTIVES.

History of Civilization.—Guizot.
International Law.
Italian.—Grandgent's Italian Composition ; Readings from Dante, Manzoni Pellico.
French.
German.
English.
Thesis in Science.

SCIENTIFIC FACULTY OF 1897-98.

The scientific faculty for the current year includes six men, besides electrician, engineer, instructors in the academic department, and janitors.

*JOHN L. GILPATRICK, A.M., Ph.D , Benjamin Barney Professor of Mathematics, was born in Granger, New York, January 12, 1845 He received his early education in the common schools of Granger and at the age of thirteen moved to Ohio with his parents. He graduated with the degree of B A., in 1867, from Kalamazoo, Mich , as valedictorian of his class. After teaching one year in the country schools he became superintendent of the public schools of Ft. Dodge, Iowa, Gos port, Ind., and Bowling Green, Ohio. He was Instructor in Mathematics in the University of Michigan from 1873 to 1874. Since 1874 he has occupied his present position in the chair of Mathematics. He is at present the senior member of the University Faculty and was for many years the University Treasurer. He is President of the Society of Civil Engineers of Ohio.

The work in his department embraces the following subjects:

Algebra An advanced course. *Geometry*. *Trigonometry*—Plane and *Spherical*. *General Geometry and Calculus*.

Instruction in Civil Engineering is given by actual field practice in Land Surveying, in Laying out Roads and Railroads, and in Leveling. Johnson's Plane Surveying and Henck's Field Book for Engineers are the text books used. The University is supplied with good instru ments for field work.

A course in *Descriptive Geometry* is open to those who have had Elementary Algebra, Elementary Mechanical Drawing, and Plane and Spherical Geometry.

---

*The department of mathematics, pure and applied, is not now located in Barney Memorial Hall, but occupies several large and commodious rooms in College Hall, which were vacated by the other scientific departments when they moved to their new quarters in Memorial Hall. These rooms were remodeled and fitted up especially for the mathematical department and are well supplied with the best apparatus.

J. L. Gilpatrick, Ph. D.

A. D. Cole, A. M.

W. G. Tight, M. S.

C. Judson Herrick, M. S.

W. H. Boughton, B. S.

H. C. McNeil, B. S.

SCIENTIFIC FACULTY OF 1897-8

Principles of Mechanism—Recitations from text-book and solutions of problems in drawing room.

Analysis of Structures, Graphical and Analytical—Open to those who have had the mathematics of the Sophomore year and Mechanics.

Strength and Resistance of Material—Open to those who have had Analysis of Structures.

The work in Astronomy is at present in charge of this department. The subject as presented in Young's General Astronomy, supplemented by lectures, is offered. Moreover it is hoped that the department of Astronomy may soon be put upon an independent basis and furnished with a well equipped observatory.

It is to be hoped, also, that the department of mechanical engineering will be put upon an equal footing with the other scientific departments, by the appointment of a professor and the equipment of the necessary laboratories. The trustees have already shown an interest in the development of this work which is receiving so much attention and for which there is a real and growing demand among the students. Their efforts in this direction will certainly be appreciated by the patrons of the school and all those interested in the teaching of science in Denison University.

* * *

ALFRED DODGE COLE, A.M., Henry Chisholm Professor of Chemistry and Physics, was born at Rutland, Vermont, December 18, 1861. He received his early education in the grammar and high schools of Beverly, Mass. He entered Brown University in 1880. In 1883 he received the Howell Premium " for highest grades in Mathematics and Physics," and was also appointed to the Oratio Latina at Junior Exhibition, and the first Junior elected to Phi Beta Kappa. He graduated from Brown as valedictorian of his class, with the A.B. degree, in 1884. After spending one year in post-graduate study at Johns Hopkins University, he took charge of the work in chemistry and physics at Denison University, which position he still occupies. He was a member of the building committee of Barney Memorial Science Hall, appointed by the trustees, and as he was constantly on the ground the great burden of inspecting the details of construction fell to his hands. This work he discharged with very great credit to himself and profit to the University.

During the school year of 1884-5 he pursued special studies in Physics in the University of Berlin. He is a member of the American Association for the Advancement of Science. Besides his publications in the Bulletin of the Scientific Laboratories of Denison University, given in the table of contents of that publication, he is the author of " *On the Refractive Index and the Reflecting Power of Water and Alcohol for Electrical Waves,*" published in Annalen der Physik und Chemie, and the Physical Review; " *Electrical Waves in Parallel Wires,*" in the Proceedings of the A. A. A. S and the Electrical World; " *Denson Univ. Laboratory Course in Electricity and Magnetism,*" and others.

More than ten thousand dollars have recently been spent in equipping the laboratories of Physics and Chemistry for efficient work. They occupy fifteen rooms in Barney Memorial Hall.

The Courses in Physics include eight and one-half terms of work in Mechanics, Sound, Electricity, Heat, Light and Electrical Engineering. The work is largely laboratory work, and the laboratory is well supplied with modern apparatus and reference books. Five large power dynamos and motors, as many more small ones, ammeters, voltmeters, spectrometers, photometers, polariscopes, etc. of recent construction are available, and electrical current for light and power is furnished by a storage battery capable of furnishing nine horsepower. A shop, well furnished with power-driven machines for work in wood and metal, furnishes opportunity for construction of apparatus for special purposes.

The course in Chemistry includes ten terms of work in General Chemistry, Qualitative Analysis, Organic Chemistry, Assaying, Sanitary Chemistry, Electro-Chemistry and Technological Chemistry. Six analytical balances are provided for this work, also spectroscope, polariscope, storage battery for electrolytic work, three assay furnaces for testing ores, Beckmann's apparatus for determining molecular weights, Hempel's for gas analysis, etc. There are working desks, well supplied with gas and water, for sixty students, and a considerable collection of reference books.

Instruction in Chemistry is given by daily lectures and recitations during the spring to Freshman pursuing the course leading to the degree of Bachelor of Arts. Remsen's Chemistry and the Laboratory Manual of the same author are text-books used. Thorough experimental illustration in the class room is supplemented by individual work in the laboratory. Abundant apparatus and desk room, with water

ADVANCED PHYSICAL LABORATORY

and gas at each desk enable each student to verify for himself, experimentally, the fundamental facts of the science. Scientific development is secured by making demonstrated facts anticipate the theoretical treatment of the subject.

The study of qualitative and quantitative analysis, required in the course leading to the degree of Bachelor of Science and elective in other courses, includes laboratory work, three days a week during one year, weekly recitations on the chemistry of the metals during one term, and weekly recitations and discussions of methods in analysis throughout the course. The use of the spectroscope is taught. Both gravimetric and volumetric methods are used in quantitative work. Appleton, Thorpe, Caldwell and Fresenius are the authors most consulted in this department of work.

Scientific Freshmen and Classical Juniors (elective) have organic Chemistry three times a week in the Winter Term.

An elective course in Water Analysis or Assaying is offered in the Fall Term of the Junior Year, and later a term each in Advanced Organic, Advanced Quantitative Analysis and Technological Chemistry. Assay furnaces, combustion furnaces, Hempel's apparatus for gas analysis, etc., are available for this work.

In Physics instruction is given to the Junior class in Mechanics and Acoustics daily during the fall term and in Magnetism, Electricity, Heat and Light during the remainder of the year. Two hours a week are occupied wholly with class room exposition, experiment and recitation; three exercise are devoted chiefly to laboratory work. The laboratory experiments are chiefly quantitative, illustrating the principal methods employed in physical research. Detailed reports of the laboratory work are prepared by the students and handed in for criticism. These form the basis for occasional talks upon laboratory methods. Students are encouraged to devise and construct apparatus, and a machine shop equipped with two steam engines, lathes, dynamos, electric motors, etc., furnishes abundant means for such work. A regular class in apparatus construction is usually formed. and much useful apparatus has been made by these classes. Two and one-half terms of advanced laboratory work in the Junior and Senior years, and one of Electrical Engineering in the Senior year are offered as electives in the [P] Bachelor of Science course. Apparatus for the accurate measurement of physical quantities is being constantly secured, and continued effort will be made to provide instruments for accurate work. Pickering,

Kohlrausch, Stewart and Gee, Thompson, Sabine and Nichols, are the authors most consulted to supplement the laboratory guide of the professor in charge. Carefully prepared reading lists give ready access to the literature of special topics.

* * *

WILLIAM G. TIGHT, M.S., Professor of Geology and Botany, was born March 12, 1865, at Granville, Ohio, where he received his education in the public school, preparatory to entering Denison in 1881. He graduated with the degree of B.S. in 1886, having devoted especial attention to science. He received his M.S. degree from Denison in 1887, and received appointment as Instructor in Science in the academic department. In 1889-92 he occupied the position of Assistant Professor of Geology and Biology and had full charge of the work of the departments. During the summer term of 1888 and the winter term of 1893 he pursued special studies in Harvard University. He is a member of the American Association for the Advancement of Science, and The Geological Society of America. Also President of the Ohio State Academy of Science and Permanent Secretary of the Denison Scientific Association. He has been editor of the Bulletin of the Scientific Laboratories of Denison University since 1889. His numerous contributions to scientific literature have mostly appeared in the Bulletin, the titles of which will be found in the tables of contents which are given further on in this volume.

He is an amateur photographer of years experience and has charge of the photo-engraving department. All of the cuts used in illustration of this volume and most of those of the Bulletin of Scientific Laboratories and Journal of Comparative Neurology and the other University publications are made by him in the department of photography and engraving.

The Department of Geology occupies several large laboratories in Barney Memorial Hall. A good equipment of lithological lathes, microscopes, models, maps and other apparatus is furnished.

The department library contains several hundred volumes and a large collection of recent literature.

The courses of instruction include Physical Geography, Structural and Dynamical Geology, Paleontology, Lithology, Mineralogy, Physiography, and Economic Geology. Special attention is given to laboratory and field work.

BOTANICAL LABORATORY

A large museum containing type forms of fossils, suites of sedimentary, igneous and metamorphic rocks, and illustrative material in dynamical and structural geology forms an important part of the equipment.

The Bulletin of the Scientific Laboratories, published under the auspices of the Denison Scientific Association, furnishes means for publication of original work.

The department of Botany which occupies several well equipped laboratories and includes a large herbarium is at present under the charge of the department of Geology.

In Geology.—In the spring term of the Freshman year scientific students begin the study of Dynamical and Structural Geology. Dana's Manual of Geology is used in the text work, which is supplemented by lectures, laboratory and field work. This is followed by a term's work in Determinative Mineralogy. The work is largely confined to the laboratory, and embraces blow-pipe analysis, the elements of crystallography, and economic mineralogy. Dana's Manual and Brush's Determinative Mineralogy, with other reference books, are used.

In the winter term Junior year geological students are given a course in physiographic geology which includes principally topographic work.

In the fall term of the Senior year Historical Geology is studied. Dana's Manual is used as a text and the student devotes much time to field work and the solving of assigned problems of local geology.

In the winter term a course in Applied and General Geology varies with the exigencies arising. The course usually embraces the study of lithology, and the application of geology to the arts. Stratified rocks are studied with reference to their microscopic peculiarities and economic application. Metamorphic and igneous species are then studied by means of thin sections and the polarizing microscope. The text books employed are Rutley, Rosenbusch, and Hussack's Tables. Laboratory practice in Paleontology is sometimes substituted.

In the spring term a course in field geology includes the solution of original problems in local geology.

The Classical student may elect a course in General Geology in the fall term of the Senior year.

In Botany.—In the fall term, Junior year there is offered an elective term of structural Botany, which includes a study of the histology of vegetable tissue. In the winter term, Junior year, a couse in Bacteriology may be elected. Sternberg's Manual forms a basis for the term's work, which consists largely of laboratory practice.

In the spring term, Junior year, a general course in Cryptogamic botany includes the study of types and is mostly microscopic laboratory work.

In the spring term there is offered also, as an elective, a course in general plant physiology and chemistry.

Standard texts are used in all the work and the botanical laboratory is well supplied with reference works of highest authority.

*   *   *

C. JUDSON HERRICK, M.S , Assistant Professor of Zoology, was born Oct. 6, 1868, at Minneapolis, Minn., where he received his elementary schooling. In 1885 he entered the Preparatory Department of Denison University, continuing his studies at this place nearly through his Sophomore year. He then entered the University of Cincinnati, from which he took the degree of Bachelor of Science in 1891.

He was appointed Instructor in Natural Sciences in Granville Academy, 1891-2 ; Professor of Natural Sciences, Ottawa University, 1892-3 ; Fellow in Biology in Denison University, 1893-5 ; Instructor in Biology, Denison University, 1895-6; University Scholar in Biology, Columbia University, 1896-7 ; Associate in Comparative Neurology, Pathological Institute of the New York State Commission in Lunacy, 1897–; Assistant Professor of Zoology, Denison University, 1897-. He took the degree of M.S. from Denison University in 1895.

Exclusive of papers in the Journal of Comparative Neurology and the University Bulletin, elsewhere noted, he has written the following articles :

Résumé of Recent Advances in the Study of the Nervous System, Transactions of the Kansas Academy of Science for 1892.

The Correlation between Specific Diversity and Individual Varibility, as Illustrated by the Eye-muscle Nerves of the Amphibia. Proc. 7th. Annual Session Assoc. American Anatomists, 1895.

Nature Studies as a Preparation for Advanced Work in Science. Ohio Educational Monthly, Vol. XLVI, No. 4, April, 1897.

The Cranial Nerve Components of Teleosts. Anatomischer Anzeiger, Bd. XIII, No. 16, 1897.

Report upon a Series of Experiments with the Weigert Method as applied to Fish Tissues. New York State Hospitals Bulletin, 1898.

HISTOLOGICAL LABORATORY

With C. L. Herrick. Articles in the Baldwin Dictionary of Philosophy and Psychology. [In preparation.]

In Zoology the preparation required is such as is usually afforded in high and preparatory schools, including an elementary course in Physiology and Hygiene, and, for scientific students, a second term in Human Anatomy and Physiology, and a term's work in Botany.

In the Sophomore year the winter term is devoted to Vertebrate Zoology, the work consisting of lectures and recitations on the structure and classification of vertebrates supplemented by. demonstrations and dissections in the laboratory. The course is intended as a general introduction to the following courses in Zoology and Paleontology. Classical students use Martin's '' Human Body '' during the corresponding term. In the spring term scientific students take up the practical study of the invertebrates, the laboratory course being accompanied by lectures on classification and the more fundamental biological problems.

The biological section of the scientific Juniors devotes the fall term to the Comparative Anatomy and Histology of vertebrates, especial attention being paid to the cultivation of the most recent methods in the microscopical examination of tissues. The course in Neurology offered to the Juniors in the winter term aims not only to impart a thorough knowledge of the anatomy and physiology of the nervous system, but to develop some of the practical hygienic and pedagogical applications. The student is assisted in the independent use of literature and introduced to the methods of biological research as applied to the morphological and practical problems of Neurology. In the spring term the same students take up Elementary Embryology, especial attention being given to problems of histogenesis and the functions of the cell in health and disease.

In the winter term of the Senior year, a course in Physiological Psychology is required of biological scientific students and is elective for others. Students expecting to take this course are strongly advised to take the Junior Neurology. After a course of lectures outlining the field, much of the time is devoted to a laboratory study of special topics, such as sensation, perception, attention, choice, the expression of emotion, etc. Kymograph, chronoscope and other necessary apparatus are supplied and well equipped machine shops give opportunity for the construction of additional pieces.

The courses in Biology are designed to bring the student face to face with nature and to encourage independence and originality of thought. The laboratories are well equipped with compound microscopes and microtomes of modern construction, together with incubators and other necessary adjuncts for instruction and research.

The "Journal of Comparative Neurology," now in its eighth volume, is published quarterly from the department of Zoology, and affords an avenue of publication for the researches conducted in the neurological laboratory.

* * *

WILL H. BOUGHTON, C.E., Instructor in Mathematics and Natural Science, was born at Bowling Green, O., May 24th, 1867. He received his early education in the public schools of Norwalk, Ohio.

From 1889 to 1891 he attended Denison University. In 1893 he graduated with the degree B.S. from the Civil Engineering Course of the University of Michigan and has since earned the second degree of Civil Engineer by graduate work in the same institution.

In engineering work he has been employed as Assistant Engineer Maintenance of Way for the C. C. C. & St. L. Ry., St. Louis Division; Structural Draftsman for the New Jersey Steel and Iron Company, Trenton, N. J.; Pittsburg Bridge Company, Pittsburg, Pa.; Brown Hoisting and Conveying Machine Company, Cleveland, Ohio. In 1894 he was elected to his present position as Assistant in the department of Mathematics under Professor J. L. Gilpatrick.

* * *

COLVER H. McNEIL, Instructor in Chemistry and Physics was born October 2d, 1866, near Winchester, Adams County, Ohio. He attended country school and completed the High School course at Winchester. After spending some months in the study of anatomy and physiology with a physician, he taught for the next five years in the public schools and in North Liberty Academy. Two summers were spent in attendance upon a Teachers' Normal School. In 1893 he entered Denison University and graduated with the degree of B.S. in 1896. During the summer of that year he pursued special work in chemistry at Harvard University and in the fall entered upon his duties as assistant to Professor A. D. Cole in Chemistry and Physics.

ENGINE AND DYNAMO ROOM OF THE POWER AND LIGHT PLANT

THE DENISON SCIENTIFIC ASSOCIATION.

The Denison Scientific Association was organized April 16th, 1887 by Professor C. L. Herrick. By his invitation a number of the professors and students met in his recitation room in College Hall. Professor Herrick presented to the meeting the plans for organization. A committee on constitution and by-laws was appointed and later reported the following :

CONSTITUTION AND BY-LAWS OF DENISON SCIENTIFIC ASSOCIATION.

CONSTITUTION.

ARTICLE I.

NAME.

This Society shall be called "The Denison Scientific Association."

ARTICLE II.

AIMS.

(a)   To afford opportunity for the interchange of ideas by those interested in the various sciences.

(b)   To collect, record and disseminate information bearing on the sciences.

(c)   To stimulate interest in local natural history and preserve specimens illustrating the same.

ARTICLE III.

OFFICERS.

The officers shall be —

First—The President, who shall preside at all meeting and exercise the powers vested in the presiding officer.

Second—The Vice President, who shall preside in the absence of the president.

Third—The Recording Secretary, who shall preserve records of all business transacted and keep such records of proceedings and communications as the Association shall vote.

Fourth —The Treasurer, who shall collect and hold in trust all moneys belonging to the Association and pay the same only upon vote of the Association, receiving a written order signed by the president and secretary.

Fifth — The Permanent Secretary, who shall be an instructor in some branch of science in Denison University, and who shall be acting corresponding secretary and ex-officio curator and special committee on publication. The permanent secretary shall hold office subject only to resignation.

Sixth —The Executive Committee. The officers of the Association shall constitute an executive committee, which shall have power to act in the name of the Association in the absence of specific instructions.

Seventh —Sections   It shall be rulable for the president to announce, at the beginning of each year, sections of the Association charged with the special superintendence of certain lines of investigation. The chairmen of these sections may be held responsible for periodical reports on the subject assigned. The sections shall include the the following subjects: 1st, Geology and Mineralogy; 2d, Biology and Microscopy; 3th, Chemistry, Physics and Astronomy; 4th, Philology, Ethnology and Explorations.

ARTICLE IV.

TERMS OF OFFICE.

All officers, except the permanent secretary, shall be elected at the first regular meeting after the opening of the fall term of Denison University.

ARTICLE V.

MEMBERSHIP.

Students of Denison University shall become members by signing the Constitution and paying the regular dues. All other candidates for membership must have their names proposed by two active members and be duly elected by a majority vote of those present at some regular meeting. Members of the Faculty shall be ex-officio members, but shall be entitled to the privileges of active membership, only upon payment of fees and dues. No persons shall be elected honorary members, except on their request or expressed permission. All active and honorary members become entitled to receive the annual publications the Association, and to the use of its library.

BY-LAWS.

ARTICLE I.

MEETINGS.

The Association shall meet for the election of officers on the first Saturday evening of the school year and shall hold bi-weekly meetings regularly thereafter during term time, at which any routine business may be transacted. Meetings may be called at the request of five members and due notification, at which meetings any business, aside from the election of officers or a motion to amend the Constitution, may be in order. Special meetings may be held, subject to the call of the president; but no business of record may be transacted.

ARTICLE II.

FEES.

All active members shall pay a fee of $.50 for one term, or $1.00 for one year, or $3.00 for four years, or $25 00 for life membership; and it it shall not be permissible to levy any other tax or assessment upon the members, except by mutual consent of all concerned.

ARTICLE III.

ORDER OF BUSINESS.

The usual order of exercises shall be :

1. Roll Call.
2. Reading of Minutes.
3. Proposals for Membership.
4. Reports of Committees.
5. Unfinished Business.
6. New Business.
7. Secretary's Report on Communications.
8. Reports of Chairmen of Sections.
9. Regular Program.
10. Informal Discussion.

Election of officers shall be construed as unfinished business.

The report of the committee was at once adopted and the organization effected with the following as charter members :

C. L. Herrick, W. H. Johnson, A. D. Cole, J. L. Deming, C. P. Jones, E. H. Castle, H. L. Jones, J. E. Woodland, D. E. Munro, W. G. Tight, T. A. Jones, A. T. VonShulz, W. E. Castle, C. L. Payne, W. H. Herrick, C. Judson Herrick, Aug. F. Foerste, E. S. Clark, C. R.

Hervey, E. A. Deming, W. H. Cathcart, Geo. D. Shepardson, Chas. Chandler, L. E. Akins, John Thorne, Enoch J. Price, Chas. T. Atwell.

As the Association increased in its membership and work it was found necessary to add two new sections to the original list so that at present the sections of the association are :

1st, Geology and Paleontology; 2d, Photography; 3d, Biology and Microscopy; 4th, Chemistry and Mineralogy ; 5th, Physics and Astronomy ; 6th, Philology, Ethnology and Explorations; 7th, Pure and Applied Mathematics. Each section leader is responsible for the program of the meeting at which his section has the principal papers. It is customary also for each section leader to make a brief report at each meeting of the progress in his department for the two weeks preceeding. In this way a résumé of the scientific literature in every department for the interim between meetings is reported at each meeting, while each section in succession is represented by more extended reports and original papers. The benefits thus gained are very great. The Association stimulates work in every department and its members are kept in touch with work being done in other lines than their own, thus acquiring a large amount of general knowledge. Before the organization of the Association Professor C. L. Herrick had begun the publication of the Bulletin of the Scientific Laboratories and as permanent secretary of the Association he also acted as editor of the Bulletin, which he placed in the hands of the Association as its official organ of publication. Thus it was that the Bulletin of the Laboratories passed under the auspices of the Association with its permanent secretary as editor. Professor Herrick remained permanent secretary of the Association until 1889, when Professor Tight was elected to the position made vacant by Professor Herrick's absence from the University.

MINERALOGICAL LABORATORY

SCIENTIFIC PUBLICATIONS.

Closely connected with the work of instruction in the class-room and laboratory is the work of investigation and research. This is being done by both professors and students in the various departments. Pupils are especially encouraged in this kind of work, as it is believed that in no other way can the powers of independent thought be so well developed in the student. The results obtained are of value also to science at large and the means for publication at present offered are the Bulletin of the Scientific Laboratories of Denison University and the Journal of Comparative Neurology.

THE BULLETIN, as already stated, was founded by Professor C. L. Herrick and the first volume appeared in December, 1885. After the organization of the Denison Scientific Association in 1887, Professor Herrick made the Bulletin the official organ of the Association, but he continued as its editor until 1889. Since that time it has been edited by Professor Tight. The work of the Bulletin can best be shown by the list of articles which have appeared in it, which is given by volumes in the following tables of contents:

VOLUME I.

With 15 Plates 5 Cuts.

VOLUME II.

With 15 Plates and 2 Cuts.

PART I.

SPECIAL SHOP OF BIOLOGICAL DEPARTMENT

Vol. X will include, besides the matter herein contained, a supplement consisting of a complete index to the first ten volumes. It has been decided hereafter to publish the Bulletin by articles, making a volume to include as nearly 500 pages as is convenient and to close with an index. The advantages of this method are believed to outweigh those of the present form of publication.

Besides furnishing a ready and convenient means of publication, the Bulletin is also of great value to the Scientific Association and the University through the large amount of valuable scientific literature which is received in exchange and part of which could be secured in no other way. It is not possible in the limits of this article to give any adequate list of the material which has been added already to our library in this way. Some idea can be gained however from the exchange list. This will show also the wide distribution which the Bulletin receives Most of these organizations favor us with their entire list of publications, often consisting of several series.

EXCHANGE LIST OF THE BULLETIN.

AMERICA (NORTH.)

UNITED STATES.

ALBANY, N. Y.,
State Library.

AMHERST, MASS.,
Amherst College, Scientific Department.

BALTIMORE, MD.,
Biological Laboratories, Johns Hopkins University.

BOSTON, MASS.,
  American Academy of Arts and Sciences.
  Boston Society of Natural History.
  Journal of American Ethnology and Archæology.
  Public Library.

BLOOMINGTON, IND.,
  Indiana University.

BROOKVILLE, IND.,
  Brookville Society of Natural History.

CAMBRIDGE, MASS.,
  Museum of Comparative Zoology.
  Peabody Museum of American Archæology aud Ethnology.

CHICAGO, ILL.,
  American Antiquarian.
  University of Chicago.

CHAPEL HILL, N. C.,
  Elisha Mitchell Scientific Society.

COLUMBUS, O.,
  Meterological Bureau.
  Geological Survey.
  Horticultural Society.

CINCINNATI, O.,
  Society of Natural History.

COLUMBIA, MO.,
  State University.

CRAWFORDSVILLE, IND.,
  Botanical Gazette.

DAVENPORT, IA.,
  Davenport Academy of Natural Science.

FORT COLLINS, CAL.,
  State Agricultural College.

IOWA CITY, IA.,
  State University Laboratories.

INDIANAPOLIS, IND.,
  Geological Survey.

ITHACA, N. Y.,
  Cornell University.

LINCOLN, NEB.,
Nebraska University.

LAWRENCE, KAS.
Kansas University Quarterly.

MADISON, WIS.,
Natural History Society of Wisconsin.
State University of Wisconsin.
Wisconsin Academy of Arts, Science, and Letters.

MERIDEN, CONN.,
Scientific Association.

MILWAUKEE, WIS.,
Public Museum.
Natural History Society.

MINNEAPOLIS, MINN.,
Academy of Natural Science.
American Geologist.
Geological and Natural History Survey.
University of Minnesota.

NEW HAVEN, CONN.,
Connecticut Academy of Arts and Sciences.

NEW YORK, N. Y.,
New York Botanical Garden.
Academy of Science.
Public Library.
American Museum of Natural History.
Microscopical Society.
Torry Botanical Club.

OBERLIN, OHIO,
Bulletin of Oberlin Laboratories.

PHILADELPHIA, PENN.,
Academy of Natural Science.
American Philosophical Society.

ROCHESTER, N. Y.,
Academy of Science.
University of Rochester.

SAN FRANCISCO, CAL.,
California Academy of Science.

SALEM, MASS.,
    Essex Institute.
    Peabody Academy of Science.
SANTA BARBARA, CAL.,
    Society of Natural History.
SPRINGFIELD, MASS.,
    Historical Library.
SPRINGFIELD, ILL.,
    Geological Survey.
ST. LOUIS, MO.,
    Academy of Science.
    Missouri Botanical Gardens.
TOPEKA, KAS.,
    Kansas Academy of Science.
    Washburn Laboratories of Natural History.
TRENTON, N. J.,
    Trenton Natural History Society.
TUSCALOOSA, ALA.,
    Geological Survey.
URBANA, ILL.,
    State Laboratory of Natural History.
WASHINGTON, D. C.,
    Geological Survey.
    National Museum.
    National Academy of Science.
    Philosophical Society.
    Department of Agriculture.
    Smithsonian Institution.
    Library Surgeon General's Office.

CANADA.

HALIFAX (*Nova Scotia*,)
    Nova Scotia Institute of Science.
MONTREAL (*Quebec*,)
    The Canadian Record of Science.
    Natural History Society.
    Royal Society of Canada.

Toronto (*Ontario,*)
    Canadian Institute.

MEXICO.

Mexico,
    Sociedad Mexicana de Historia Natural (*Mexican Natural History Society.*)
    Sociedad de Geographia y Estadistiea de la Republica Mexicana.

AMERICA (SOUTH).

ARGENTINE REPUBLIC.

Buenos Aires,
    Museo Nacional Annals (*Annals of National Museum.*)
    Sociedad Cientifica Argentina (*Argentine Scientific Society*)

BRAZIL.

Rio de Janeiro,
    Museo Nacional (*National Museum.*)

CHILE.

Santiago,
    The University.

URUGUAY.

Montivideo,
    Museum of Montivideo.

ASIA.

JAPAN.

Tokio,
    Tokio Daigaku (Imperial University.)

AUSTRALIA.

NEW SOUTH WALES.

Sidney,
    Geological Survey.
    Linnean Society of New South Wales.
    Australian Museum.

SOUTH AUSTRALIA.

Adelaide,
    Royal Society of South Australia.

VICTORIA.

Melbourne,
    Royal Society of Victoria.

EUROPE.

AUSTRIA-HUNGARY.

GRATZ (*Styria*,)

Der Naturwissenchaftliche Verein für Steirmark (*Society of Natural Science.*)

PRAGUE (*Bohemia*,)

Kaiserliche böhmische Gesellschaft der Wissenschaften (*Royal Bohemian Society of Science.*)

VIENNA (*Austria*,)

Kaiserliche Akademie der Wissenchaften (*Imperial Academy of Science.*)

BELGIUM.

BRUSSELS,

Académie Royale des Sciences, des Lettres et des Beaux Arts de Belgique (*Royal Academy of Sciences, Letters and Fine Arts of Belgium.*)

Musée Royal d'Historie Naturelle de Belgique (*Royal Museum of Natural History of Belgium.*)

DENMARK.

COPENHAGEN,

Kongelige Danske Videnskabernes Selskab (*Royal Danish Society of Sciences.*)

FRANCE.

ANGERS,

Société d'Etudes Scientifiques (*Society of Scientific Studies.*)

CHERBOURG (*Manche*,)

Société Nationale des Sciences Naturelles et Mathematiques de Cherbourg (*Society of Natural Sciences and Mathematics of Cherbourg.*)

NANCY,

Société des Sciences de Nancy (*Society of Sciences of Nancy*)

PARIS,

Société Zoologique de France (*Zoological Society of France.*)

Société Geologique de France (*Geological Society of France.*)

Société Anatomique de Paris (*Anatomical Society of Paris.*)

ROUEN,

La Société des Amis des Sciences Naturelles (*The Society of the Friends of Natural Science.*)

SHOP

GERMANY.

BERLIN (*Prussia*,)

Königliche (Preussiche) Akademie der Wissenchaften (*Royal Prussian Academy of Sciences.*)

BREMEN (*Germany*,)

Naturwissenschaftlicher Verein (*Society of Natural Sciences.*)

DRESDEN (*Saxony*,)

Gesellschaft für Natur- und Heil-kunde (*Society of Natural and Medical Sciences.*)

FRANKFORT-AM-MAIN,

Senckenbergische Naturforschende Gesellchaft (*Senckenberg Naturalists' Society.*)

GÖTTENGEN (*Prussia*,)

Königliche Gessellschaft der Wissenschaften (*Royal Society of Sciences.*)

HALLE-AN-DER-SAALE (*Prussia*,)

Kaiserliche Leopoldini Carolina Akademie der Deutschen Naturforscher (*Imperial Leopold-Carolus Academy of German Naturalists.*)

Naturforschende Gesellschaft (*Naturalists' Society.*)

HEIDELBERG (*Baden*,)

Naturhistorisch-Medicinischer Verein (*Society of Natural and Medical Sciences.*)

KÖNIGSBERG-IN-PRUSSEN (*Prussia*,)

Die Königliche Physikalisch Oekonomische Gessellschaft (*Royal Physico-Economical Society.*)

MUNICH (*Bavaria*,)

Königlich Baierische Akademie der Wissenschaften (*Royal Bavarian Academy of Sciences.*)

GREAT BRITAIN AND IRELAND.

ENGLAND.

BRISTOL,

Naturalist Society.

BATH,

Journal of Microscopical Society.

CAMBRIDGE,
 Philosophical Society.
LONDON,
 Biblographical Bulletin.
 Geological Survey of the United Kingdom.
 Patent Office Library.
 Society for Psychical Research.
MANCHESTER,
 Literary and Philosophical Society.

IRELAND.

DUBLIN,
 Royal Dublin Society.
 Royal Irish Society.

SCOTLAND.

EDINBURGH,
 Royal Society of Edinburgh.
GLASGOW,
 Natural History Society of Glasgow.

ITALY.

NAPLES,
 Reale Academie delle Scienze Fisiche e Mathematiche (*Royal Academy of the Physical and Mathematical Sciences.*)
 Sociéta Reale di Napoli (*Royal Society of Naples.*)
LUGANO,
 Société Helvetique des Sciences Naturelles (*Helvetian Society of Natural Sciences.*)
ROME,
 Officio Geologico (*Geological Office.*)
SOLOTHURN,
 Naturforschende Gesellschaft (*Society of Naturalists.*)
TURIN,
 R. Academia della Scienza (*Royal Academy of Science.*)
 Musei di Zoologia ed Anatomia Comparata delle R. Università di Torino (*Museum of Zoology and Comparative Anatomy of the Royal University*)

NORWAY.

CHRISTIANA,
Norske Geologiske Undersaegelse (*Norwegian Geological Survey.*)

RUSSIA.

Moscow,
Imp. Moskofskoie Obshchestyo-Ispytatetei (*Imperial Society of Naturalists of Moscow.*)

RIGA,
Obschchestyo Iestestyo-Ispytateleie (*Society of Naturalists.*)

ST. PETERSBURG,
Comiti Geologique (*Geological Survey.*)
Institute des Mines (*Institute of the Mines.*)
Russisch-Kaiserlichen Minerologischen Gessellschaft (*Royal Mineralogical Society.*)

KIEW,
Société des Naturalistes (*Society of Naturalists.*)

SPAIN.

MADRID,
Real Academia de Sciencias de Madrid (*Royal Academy of Sciences of Madrid.*)

SWEDEN.

LUND,
Acta Universitatis Lundensis (*Royal University of Lund.*)

STOCKHOLM,
Kongligi Svenski Vetenskaps Akademien (*Royal Swedish Academy of Sciences.*)
Sveriges Geologiska Undevsökning (*Swedish Geological Institute.*)

UPSALA,
Kongliga Vetenskaps Societaten (*Royal Society of Sciences.*)

SWITZERLAND.

BASEL,
Naturforschende Gesellschaft (*Society of Naturalists.*)

BERN,
Naturforschende Gesellschaft (*Society of Naturalists.*)

THE JOURNAL OF COMPARATIVE NEUROLOGY was founded in 1891 by C. L. Herrick, then Professor of Biology in the University of Cincinnati. It is devoted to the study of the nervous system of man and the lower animals so far as these studies are conducted by the comparative method in the broadest sense of that term, a field not occupied by any other periodical in any language. Its pages are chiefly occupied with original contributions to science and these are fully illustrated with many plates. An important feature, however, is the review department by means of which the reader is kept in touch with the latest and best current literature both in this country and abroad. Due attention is given to the psychological bearing of neurological discoveries and to the data of comparative psychology.

In 1892, upon Professor Herrick's return to Denison University, the Journal became one of the university publications and since then an annual grant has been made for its support. In 1894, Professor Herrick's brother, C. Judson Herrick, became one of the editors and has since that time been the business manager. In 1896 the editorial staff was reorganized with Professor C. L. Herrick as Editor-in-chief and Oliver S. Strong of Columbia University and C. Judson Herrick of Denison University as Associate Editors. In 1898, beginning with the first number of Vol. VIII, there have been added to the editorial staff a number of Collaborators, all men of eminence in their respective departments and representing leading institutions in this country and Europe. At the present writing the list of Collaborators is not yet completed, but the following may now be announced :

Henry H. Donaldson, Ph.D., *Professor of Neurology, University of Chicago ;* Growth and regeneration of nervous organs.

Professor Ludwig Edinger, *Frankfort, a M.,* Collaborator for Germany.

Professor A. van Gehuchten, *University of Louvain, Belgium ;* Collaborator for France and Belgium.

G. Carl Huber, M.D., *Assistant Professor of Histology and Embryology in the University of Michigan ;* The sympathetic system and the peripheral nervous system.

B. F. Kingsbury, Ph.D., *Instructor in Microscopy, Histology and Embryology, Cornell University and the New York State Veterinary College ;* Morphology of the lower vertebrates (Ichthyopsida).

Frederic S. Lee, Ph.D., *Adjunct Professor and Demonstrator of*

ADVANCED CHEMICAL LABORATORY

Physiology, College of Physicians and Surgeons, New York City; Physiology of the nervous system.

Adolf Meyer, M.D., *Docent in Psychiatry, Clark University, and Assistant Physician to the Worcester Lunatic Hospital;* Human neurology.

The exchange list of the Journal includes may publications not received by the Bulletin.

The original articles which have appeared in the seven volumes now completed are indicated in the Table of Contents below.

ORGANIC CHEMICAL LABORATORY

BULLETIN OF THE LABORATORIES

NORTH FRONT BARNEY MEMORIAL SCIENCE HALL

THE GIFT OF BARNEY MEMORIAL SCIENCE HALL.

Early in the nineties, it became evident that the pioneer labors of Professor Osbun of the department of Physics and Chemistry and Professor C. L. Herrick of the department of Natural History, were bringing forth abundant fruit and that the scientific departments of the University were displaying a new life and energy. This was seen in many things, but especially in the large number of students who sought the courses in science. Hitherto the number of scientific students was relatively small. Now they were increasing on every hand. Indeed, so rapid was this growth that it overtaxed the equipment and became a source of embarrassment to the professors. For a certain laboratory capable of accommodating twenty students, there were fifty applicants. Every thing became over-crowded and the necessity for a new Science Hall became apparent. The matter was laid before the Trustees and a committee was appointed to solicit funds for the erection of a suitable building. This committee did some valuable work, but results were not readily reached.

The commencement of 1892 drew near, and still there was no provision for the building. Early in May of this year, the president of the University presented the facts of the case to Mr. E. J. Barney, of Dayton, and asked his advice. He at once advised that a competent architect be sought and preliminary sketches of such a building as was needed be prepared and presented to the Trustees at their approaching meeting in June. Messrs. Peters and Burns, architects, of Dayton, O., were engaged to prepare the work as suggested. This was all that was known at that time of Mr. Barney's generous designs. Commencement time came, and still there was no provision for the much needed building. At the annual meeting of the Trustees there was a general feeling of anxiety because of the situation. But Mr. Barney greatly relieved and gratified the Board by coming forward with a generous offer of $40,000 for the erection of a Science Hall. The offer was promptly accepted and a Building Committee appointed to take charge of the work of construction. All were filled with rejoicing when, two years later, the handsome and commodious building was publicly dedicated to the noble purposes for which it had been erected. During the ceremonies of dedication, Mr. Albert Thresher, of Dayton, acting on behalf of Mr. Barney and the Building Committee, presented the keys of the building, with well chosen words, to Dr. H. F. Colby, President of the Board of Trustees. Dr. Colby responded appropriately and felicitously. The dedicatory address was given by Prof. J. J. Stevenson, Ph.D., LL.D., of the University of New York, and a most eventful occasion in the history of Science at Denison University was happily closed, the only regret being that the generous donor of the building was unable to be present.

E. J. BARNEY, ESQ.

EUGENE J. BARNEY.

The donor of Science Hall, Mr. Eugene J. Barney, is the eldest
son of Eliam E. and Julia (Smith) Barney, and was born at Dayton,
Ohio, Frebruary 12, 1839. His education was received in the public
schools of his native city and at Rochester University. Soon after
leaving college he turned his attention to business. purchasing an interest in what was then known as the Barney & Smith Car Works. He
was gradually promoted through various positions of trust connected
with this large establishment and upon the death of his father in 1880
became its president. His sound judgment and ripe experience have
made him a valuable counsellor in business circles and he is officially
connected with a large number of important corporations at home and
elsewhere.

. Mr. Barney lives in an elegant home in the central portion of the
city, but still retains his membership in the Linden Avenue Baptist
Church, which he helped to establish as a mission of the First Baptist
Church in the days of his young manhood. Notwithstanding his
large business interests, he still finds time to devote to the duties of his
church and Sunday School, of which he was superintendent for many
years ; to the work of the Young Men's Christian Association and more
than one struggling young man in the city has been encouraged to right
living by the kind and thoughtful words of this Christian gentleman.

On his twenty-third birthday, Mr. Barney was married to Miss M.
Belle Huffman, eldest daughter of the late William P. Huffman, for
many years an honored trustee and beloved benefactor of Denison University; and thus the future interest of the young man in that institution
of learning seems to have been doubly assured. What lasting impressions were made upon his mind by the conferences of his father with
these friends and the elder Thresher to which he was admitted! What
loyalty was inspired by their sacrifices in the early history of the college ! What a monument is Science Hall of love toward a noble father
whose name was already inseparably linked with that history ! What

a well-spring of joy within the heart of the giver as he contemplates its broad field of usefulness! What an object lesson of consecrated wealth to every student who enters its portals! The history of Denison could not be written without mention of the elder Barney—and the Scientific Department of the future will date its real beginning and credit its real progress to this generous gift of the son.

Dignified and modest yet cordial in manner, cultured and refined from contact with books and men as well as by extensive travel, a devoted lover of home and family, a keen observer of passing events, conscientious and honorable in his business life, Mr. Barney has been an honor and a blessing to the city and community in which he lives, to the Baptist denomination in general and to Denison University in particular.

ELIAM E. BARNEY

ELIAM E. BARNEY.

*Eliam E. Barney, to whose memory Barney Memorial Science Hall was erected, was born at Henderson, New York, October 14, 1807. His father, an enterprising farmer, had made up by private effort for the deficiencies of his school training and was a man highly esteemed among his neighbors. He served in the War of 1812 and his wife, Nancy Potter of Masachusetts, was the daughter of a soldier of the Revolution. Eliam was the oldest of eleven children. Among the traits recorded of his boyhood, perhaps the most suggestive of his future career were the industry and the orderly method which he displayed in all that he had to do. In his fourteenth year he experienced conversion and became a member of the Baptist Church, a connection which he maintained during a long lifetime, bringing ever increasing wisdom into the councils of the denomination and showing towards worthy denominational enterprises a liberality commensurate with the means at his command.

His preparation for college was secured at Belleville, New York, in the school since known as Union Academy. Rev. Joshua Bradley, who fouunded the Academy, and lived for a time in the Barney household, afterward came to Ohio and acted as a traveling agent for Denison University, then known as Granville College. Eliam entered as a Sophomore at Union College, riding from his home to Schnectady, one hundred and thirty miles, in his father's wagon. To pay the expenses of his college course he had borrowed some money from an uncle and to this was added what he could earn by teaching a writing school and acting as tutor for less advanced students. During a part of his senior year he taught a public school, reciting with his class only at intervals. Thoroughness and accuracy in thought and expression was his ideal.

He was graduated in 1831, and was soon afterward made principal of Lowville Academy, a position which he filled very successfully for the two years following. Among his assistants here was Miss Julia

---

*Abridged from Memoir by Rev. H. F. Colby, D.D.

Smith, who afterwards became his wife. After teaching at Lowville long enough to pay off the debts incurred for his education, he came to Ohio, whither at his suggestion his father and family had already removed. Visiting Granville in 1833, he was asked to accept for a few months the position of Professor Drury, who had been appointed to a place in the Faculty but had not yet arrived. In this position he was known as a very popular and effective teacher.

As his temporary term of employment at Granville drew to a close, he wrote letters to the Postmasters of several towns in Ohio, inquiring for a possible opening as a teacher   The Postmaster at Dayton alone responded. He accordingly went to Dayton and found that a principal was to be chosen for the Dayton Academy. He and another applicant were called before the trustees to make a statement as to what they each considered the best method of conducting such a school, and as a result of this test he was immediately chosen. The terms proposed to him and accepted were that he should take the building rent free, assume all financial risks and receive all profits. A brother and sister were at once taken as assistants, and two other sisters were afterwards added, as the school rapidly grew in numbers. In 1838, he was an active promoter of the first movement to provide for public schools in Dayton. In 1839, the trustees of the Academy insisted that he should begin to pay rent for the use of the building, and not deeming this equitable he resigned and opened a school at his own house. This was afterwards removed to the basement of the Baptist church and continued with success for about two years when signs of failing health caused him to seek employment of a different nature. He then purchased a sawmill and managed it for two years and a half with financial success, but entered so actively into the labor of his business that his health was injured, rather than benefitted, by the change. A trip to the South, followed by another to the East, brought renewed health, and he was then persuaded to enter the educational field again, as principal of Cooper Seminary, founded in 1844. This school opened its doors in September, 1845, and enrolled one hundred and seventy-four young ladies during its first year. The Seminary made a noble record for thorough work, honestly living up to the statement of its advertisement, "Those who dislike to study or are unwilling to comply cheerfully with all school regulations, or who go to school merely because they are sent, will do well to seek some other place in which to idle their time away. We wish only such to attend as are desirous to make rapid improvement and determined to apply themselves closely to study."

Mr. Barney had taken the Seminary for five years, and at the expiration of this period decided again to go into business. He formed a partnership with Mr. Ebenezer Thresher, and after careful consideration they decided to engage in the manufacture of railroad cars. For the first year, however, Mr. Barney was a silent partner, continuing his work at the head of the Seminary that long at the urgent request of the trustees. Mr. Barney and Mr. Thresher put in $5,000 each at the beginning of their manufacturing partnership, and the business prospered from the start, their cars obtaining a reputation for excellent workmanship and material. In 1854, Mr. Thresher withdrew, because of ill-health, and was succeeded in the firm by Mr. Caleb Parker, of Massachusetts. The firm suffered temporarily from the financial panic of 1857, but was on too solid a footing to receive permanent harm. Mr. Parker retired from the business in 1864, and was succeeded by Mr. Preserved Smith. In 1867 the firm was incorporated under its present name " The Barney and Smith Manufacturing Company, of Dayton, Ohio," and its history has been one of safe and steady progress ever since. A description of the works at the present time would hardly be called for by the nature of this sketch, but perhaps it will not mislead to reprint a description written soon after Mr. Barney's death, reminding the reader that the business has made during the seventeen years intervening just such progress as might have been expected from the foundation which Mr. Barney had laid :

"The visitor to the works as they are at the present time (1881) cannot fail to be impressed by their extent and the amount of painstakng labor which is there employed. Every kind of car, from the common platform to the most luxurious drawing room or sleeper, is turned out by skillful workmen, and the rapidity with which large contracts can be filled has often occasioned surprise. The blacksmith shop with its many forges ; the large machine shop, with its complicated and beautiful appliances for working iron economically ; the foundry, that can turn out one hundred and forty wheels a day ; the two buildings of fine machinery for cabinet work; the separate shops for putting together the trucks, the freight cars, the passenger cars, or for painting them all filled with work in different stages of progress, and populous with men laboring together with exact system and precision—form a little world of industry and of wonderful interest to a thoughtful mind.

Two large engines furnish the motive power, wire ropes transmitting it to the various buildings. Such are the facilities for drying and working lumber that freight cars have been delivered, ready for use, in a few weeks only after the wood composing them was growing in the forest; but to keep available material for every kind of work, the Company has an extensive yard, with a railroad through the center, wherein is piled lumber for two hundred and fifty passenger cars and one thousand freight cars—the usual stock carried amounting to nine million feet. The factory, with its appurtenances, covers twenty eight acres of ground, and the capital stock now amounts to seven hundred and fifty thousand dollars. At Mr. Barney's death the number of employes was over a thousand."

Such, in 1881, was the industry which Mr. Barney and Mr. Thresher had founded thirty years before with a capital of only ten thousand dollars.

From the same source as the above (A Tribute to the Memory of Eliam E. Barney, by Rev. Henry F. Colby, Pastor of the First Baptist Church, Dayton, Ohio) we quote the following paragraph in description of Mr. Barney's characteristics as a business man : " Of course, he was often indebted to the prudent counsel of the partners with whom it was his privilege to be connected, and to the faithful co-operation of experienced men who were intrusted with the different departments. But for many years he was at the head of the establishment, and to his personal traits its growth and reputation were largely owing. Except when he was traveling for the business—and a large portion of his time was thus occupied, sleeping cars furnishing him his rest as he traversed wide sections of country—he was wont to be at the factory from early morning till the whistle sounded at night. He was conscientious, laborious and watchful in the extreme. He not only superintended subordinates but seemed to keep his eye with wonderful particularity on the innumerable details of the work. His presence and impress were everywhere. In the factory his was the living spirit among the wheels. He had the decision, the power, the control of an imperial commander. Each employe must come promptly up to the terms of his engagement and fulfill it; for the last hour of his day's work was the Company's profit, the other hours were necessary to earn his wages. Any form of ill behavior was reprimanded, no matter on whom the censure might fall. No work must be turned that was not the very best. Employes at first, like some other people, took his strongly marked visage, his strictness, his positive judgments and peremptory answers, for stern-ness. Sometimes his replies to those who sought his counsel would be brief, and he would seem to be absorbed in something else. He would even at times disregard those little courtesies which make men seem approachable. But if the matter really was one that required his help, the applicant would find in a day or two that Mr. Barney had thought it all over and had some plan to suggest or some relief to offer. He threw men upon their own resources to develop them, and then reached out his hand to keep them from falling. They who came to know him well found out that no one could have a kinder heart or be more ready to help those in trouble."

Doubtless many who read this bulletin have met in one place and another evidences of the deep interest which Mr. Barney took in ad-vocating the cultivation of the Catalpa tree for timber. He gathered and published statistics as to its rapid growth and durability, and through the columns of the Railway Age and other journals suggested that railroad companies should cultivate the tree along their lines for use as ties. Letters of inquiry concerning the tree came to him by thou-sands, and as the result of his efforts millions of Catalpa trees are now growing in this and other countries.

His gifts to Denison University at various times aggregated nearly seventy thousand dollars, and he was also instrumental in securing the financial support of others. In addition to this he gave much time and thought to the school as a member of its Board of trustees. Doubtless to his services as a teacher also, in 1833, is due in part the tradition of thorough work in the class room which it has always been the special effort of the University to maintain.

This hurried sketch must necessarily fail to give anyone a com-plete conception of Mr. Barney's lifework and personal characteristics. It may indicate, however, the type of men to which he belonged,— men of keen insight, unceasing industry, thorough and orderly habits of thought, conscientiously upright in every detail of life, of unobtru-sive modesty and yet strong in conviction, and always generous to any good cause or deserving individual in proportion to the means at command.

DEDICATORY EXERCISES.

The dedication of Barney Memorial Science Hall occurred June 13th, 1894. The following account of the exercises is taken from the press notice in the *Granville Times* of June 14th :

" The dedicatory exercises of Barney Memorial Hall occurred on Wednesday morning at a quarter past nine o'clock. As the cases had not yet been placed in the Museum a platform was erected in its east end for the speakers and the main floor and gallery, with all the available space in the adjoining hall, was crowded with visitors eager to participate in celebrating the completion of a building which means so much for the future of Denison in the field of scientific investigation.

" All regretted the absence of Mr. Barney, through whose liberality the building was erected, but the ceremony of turning over the keys to the Board of Trustees was fittingly performed by the Chairman of the Building Committee, Mr. Albert Thresher. Dr Colby, the President of the Board whose remarks at the laying of the corner stone are remembered by many, accepted the keys with a graceful recognition of the faithful manner in which the committee had discharged its trust, and presented them in turn to President Purinton.

" In the remarks made by Dr. Purinton special stress was placed upon the recognition of the great truths of the Christian religion in all the scientific work of the University. Attention was also called to the recognition of the work of the Denison laboratories of leading scientists throughout the civilized world as shown by the exchange of the publications of learned societies in all lands for the bulletins of our laboratories and the Journal of Neurology.

" At the close of the President's remarks an earnest and touching dedicatory prayer was offered by Rev. Mr. Lounsbury, of Dayton, pastor of the church with which Mr. Barney is connected.

" The President then introduced the principal speaker of the occasion, Prof. J. J. Stevenson, of New York.

" The leading feature of the exercises as a whole was the assurance given that Denison is not to make the vital mistake, in entering upon a broader career of scientific investigation of nature's facts and laws, of leaving out of consideration the divine mind of which these facts and laws are but the outward expression."

VIEW OF GRANVILLE FROM BARNEY MEMORIAL SCIENCE HALL

DEDICATORY ADDRESS.

SCIENCE AS AN EDUCATIONAL FACTOR.

When a magnificent building is dedicated to scientific studies, one's thoughts turn naturally to consider the bearing of such studies upon education itself: and so it has come about that Science as an Educational Factor, is the topic upon which and around which I am to speak to-day.

When patristic philosophy established a tribunal to which should be referred all questions whether of physical science or of theology, it closed the door to individual thought and opened the way to the bondage of the dark ages. This new philosophy practically proclaimed that investigation beyond what is revealed in Scripture is science falsely so-called—Augustine himself, when discussing the existence of the antipodes, said that "it is impossible that there should be inhabitants on the opposite side of the earth, for no such race is recorded in Scripture among the descendants of Adam." Necessarily, the study of nature was forbidden in fact, if not in word, physics being regarded as merely tributary to revealed theology. Monks and schoolmen occupied themselves largely in making copies of the "Fathers" or in applying the principles of Aristotelian logic to systematization of all things, physical and metaphysical. The revival of learning, though influenced by the Arabian mode of thought, carried into Europe by the Jews, was but a revival of intellectual activity along lines of study pursued for centuries. Monasteries yielded stores of ancient literature, which they had preserved as an old chest preserves valuable documents ; authors, known until then only by name or by garbled extracts, became familiar acquaintances, while to them were added hosts of others, previously unknown, whose works afforded full scope for the scholarly acumen of the time.

The Universities of the middle ages taught only such matters as engrossed the attention of the schoolmen ; disputations respecting mere abstractions occupied most of the time and absorbed most of the energy of learned men. True, the love of money and the fear of death led many to search for the philosopher's stone and for the elixir of life, but such studies lay outside of the legitimate lines and those who pursued them were viewed askance. When the revival of learning came, University courses were lengthened and broadened, it is true, but only along the old lines and within the old areas.

The distrust of physical investigation engendered in the ignorance and dread of the dark ages, when popular religion had become almost fetichism, passed away slowly. As in the later days of the gloom, weird tales were circulated respecting Friar Bacon, so even after the revival of learning, doubt pursued the investigator and those adhering to the patristic philosophy were able to thrust Copernicus and Galileo aside. It is true that in Italy, where Jewish and therefore Arabian influence had been felt very early, important studies were made ; Leonardo da Vinci and Frascatoro rediscovered the Pythagorean doctrines enunci ated by Ovid; physicists and naturalists made noble discoveries, but in great part their results lay buried and almost unstudied until the close of the last century. Even in the early part of this century, the Coper nican system was barely known to the great University of Salamanca and the works of Galileo and Copernicus remained on the Index Ex purgatorius until 1828.

Within the memory of some who are present, the terms "scientfic man" and "infidel" were, to the majority of good people, practically synonymous, though Dr. Dick's remarkable statement regarding the undevout astronomer had led many to make an exception in favor of the star-gazers. Even when a scientific man asserted his belief in rev elation, not a few doubted. It is the prerogative of ignorance to de spise or to distrust that which it cannot understand; but not unfre frequently a package, dreaded as an infernal machine, has proved to be a gift of inestimable value. So here: the dread of physical science has been disappearing rapidly of recent years, for men have come to feel that creation and revelation, having a common origin, must agree in so far as they follow parallel lines, and that disagreements are apparent, not real, being due to error in the interpretation of one or possibly of both. It is too soon to attempt a full reconciliation of the two, as we know them ; more study, more growth must be had before men can be fitted for the task ; it is still difficult to distinguish between faith and prejudice ; Scripture has been overlaid so deeply with prejudices and traditions, that we hesitate often to accept as truths the discoveries made by naturalists and archæologists ; too often, when convinced of error, we give only a half-hearted, half resentful assent to the truth which we can dispute no longer.

The imperfect recognition of the importance of scientific studies, which has continued almost until now, was due in great part to the half distrust with which all physical investigation was regarded; but another influence was almost equally potent. It was an inheritance from older days. Education, formerly, was for the wealthy, for men who were to be cloistered students or lawyers or physicians, all of them, even the physicians, dealing almost wholly in abstractions. Matters of practical utility were beneath the contempt of scholars; utilitarianism concerned only the vulgar sphere of commerce and manufactures. This conception appears absurd to us now, but not long ago, its defenders dominated our colleges, controlled the professions and moulded public opinion; the community believed that study of material things does not cultivate the intellect, that the only elevating studies are those derived from antiquity, with, as the capping stone, that pure philosophy, to which those who study gross or material things can never attain.

The importance of scientific education has been conceded in America, where recognition of the close relations between abstract and applied science could not be avoided; for the application of principles discovered by closet students has made available the mineral wealth of our vast domain, until the United States has become one of the greatest of manufacturing nations. Some Americans, who know little of what their countrymen have done in pure science, seem to regard most of the discoveries in applied science as practically piracies from European students. But Americans have made contributions second in importance to those of no other country; from the days of Franklin to our own time each generation has born its full share of burden in erecting the scientific edifice. Franklin's discovery of the identity of lightning with frictional electricity opened a new world of investigation, while leading to the protection of man against his most dreaded enemy; Rumford's investigations of heat were not understood in all their bearings for half a century, but were the suggestion for Joule and his contemporaries; Henry's studies in electricity opened the way for Morse and Vail and made the magnetic telegraph a possibility; John W. Draper's investigation of light and his investigation of the spectrum, made thirty years too soon, were the first long strides toward the development of spectrography, which, in the hands of German students, has told us of new elements and, in the hands of Young, Langley, Pickering and other Americans, has told us of the composition of our own as well as of other suns so far away as to be scarcely visible to the unaided eye; the names of Hare, Gibbs and Remsen tell us of stages in the progress of chemistry; while Newcomb, Hall, Barnard, Newton and their contemporaries have done their full share in the advance of astronomy; in ethnology are the monumental works of Pickering and Hale as well as the splendid contributions published by the United States government during the last score of years; in botany, the publications of Torrey, Gray, Englemann, Watson and a score of others are models of accuracy and beauty; in zoology, Baird, Cope, Binney, Hallowell, Marsh, Osborn, Verrill and their many co-workers have labored on the rich faunas of this great continent and their works are regarded as among the marvels of our time; in psychology, so rapidly passing from the region of mere metaphysics to the rank of an inductive science, American investigators are unexcelled even by the patient Germans; time would fail me to tell of those who have attained worldwide fame in geology since the time when Hall, Rogers and Dana were the youthful pioneers, to this day, when instead of half a score, as in 1837, we count more than two hundred and fifty active geologists. America's surveys, geological, geodetic and coast, have been the most extended in the world and the hundreds of ponderous volumes, issued by state or general government and distributed with lavish hand, have astonished other nations—as well they might.

The record of Americans in applied science is even more remarkable than that in pure science; Holley remade the whole Bessemer process so that steel rails can be made in this country for little more than one sixth of the price prevailing twenty five years ago, and our great buildings can be constructed of steel for far less than of iron; the application of Henry's curious apparatus, as made by Morse and Vail, has been modified by a score of workers until at length, by Edison's improvements, it has become not a luxury but an ordinary means of communication; the engineering feats of Americans on the railways of California, Oregon, Venezuela and Chili are unrivaled; but it is not possible to go on with such a list; our advance along all technical lines causes other nations to regard America not merely with admiration but even with perplexed wonder. On one occasion the *London Times* said:

"In the natural distribution of subjects, the history of enterprise, discovery and conquest and the growth of republics fell to America and she has dealt nobly with them. In the wider and multifarious provinces of art and science, she runs neck and neck with the mother country and is never left behind."

With all this ever present before the American community, it is not surprising that enormous gifts have been made for the foundation of scientific and technical schools; but it is surprising that the educational value of scientific training is so little appreciated and that, in so many cases, technical courses, those involving direct application of scientific priciples, are regarded as of less pedagogic value than are those which concern merely the operations of man's intellect or the immediate products of that intellect. Let us inquire for a little into the educational value of the observational sciences as well as of the technical science growing out of their applications.

Education of to-day is necessarily different from that of one or two centuries ago; then culture alone was sought, often perfunctorily, usually by the wealthier classes and with a view to one of the learned professions, then only three, law, medicine and theology; education then was for the few; now it is for all; then it was a luxury; now it is a necessary preparation for life's work; it is a training, that a man may be able to make the most of himself in some one of the now many learned professions or in some one of the complicated groups of commercial operations. But it is more than mere training, for it has two important provinces: first, to draw out and to train the mental powers; secondly, to impart knowledge. Too long, a disproportionate stress was laid on the former; there is a tendency now in many quarters to lay too great stress upon the latter. The former is the more important, but it must not be separated from the latter.

Mental faculties or powers are not independent, even in the sense that a man's limbs are independent portions of his body; the notion of this independence is but a make-shift arising in the class room. Let the mind be regarded as an entity, manifesting itself in many ways, and capable of forming habits or tendencies to act in one direction preferably to another; unguided in its formative period it will come to work along narrow paths, determined by prejudice rather than by reason. Here, as is usually the case, the intellectual powers alone are considered, for ordinary educational work has comparatively little to do directly with culture of morals, though it has much to do with it indirectly; a true culture of the intellect leads to a genuine ethical culture by inducing a judicial frame of mind, which prefers the right to the wrong.

In the normal child knowledge is acquired first by observation — through the senses; this acquisition leads to the development of that complex series, the power of retaining, that of recalling, and that of recognizing impressions, which altogether make up what is known popularly as memory. But in the process of mere acquiring, the observed things are compared and in that of recognition, things or impressions are recognized largely by their relations; this involves the examination of things apart from other things, of their differences as well as of their relations, the formation of intellectual images and the separation of essentials from non-essentials; whence the wonderful and perplexing queries with which a child assails those who can be reached; all of this leads to the formation of conclusions, of inductions, of general principles; thence to application of principles to matters not so familiar—to the formation of deductions and to the encouragement of a lively imagination.

This briefly is the succession, whether the child be of savage or of civilized parentage. How necessary guidance is during the unfolding we know only too well by observing those who have not had it. Left to itself the mind, seeing things wrongly, makes no effort to see them rightly; fails to apprehend their relations and makes inductions which are absurd and are liable to become dangerous as guiding principles of conduct. We may laugh or in better temper we may smile sadly as we read of Kaffirs who worshipped an anchor as a powerful fetich, because the man who had knocked a chip from one of the flukes, died suddenly; or we may be amused by the folly of a savage, who recognizes a demon in a gentle breeze, which, blowing on his neck, gives him a cold; but these can give a reason for their belief and conduct equally good with that which most of us can give for many beliefs influencing our action.

Education is to guide in this process of development, but creation is not within its province; cannot give intellect or common sense; but if rightly conducted it may strengthen feeble powers as gymnastic exercises may make the left hand almost equal to the right; it can take the ill-developed entity with irregular surface, not to cut away or diminish any power but to strenghten those that are feeble. As a gardner, desirous of gaining more shoots from the plant, spreads the crown that light may come within and cherish the dwarfed buds, so educational

STORAGE BATTERY ROOM

training endeavors to make the mind stronger, broader, more sym-
metrical, and, at last, finer, that the character may have at first
strength, then beauty. Success in the effort is not always certain, for
the mental treasure is in earthen vessels, very frail, too often of poor
material, very porous and without much glaze; but we have the ideal
—how may we attain at least partial success, the most possible?

Two schools answer this question; but they have little in common
beyond the belief that there is a human mind which is in sore need of
cultivation.

The old school finds the best means in the study of abstractions;
it holds that the study of languages, especially of the classical tongues,
affords the best basal training; it would place a child in earliest youth
at this study to sharpen the intellect by dwelling on niceties of expression and on the recognition of delicate distinctions, so producing exact-
ness of thought and precision of statement while strengthening the
verbal memory; with this study, though subordinate, is to be associated that of mathematics, with excursions in other directions; but em-
phasis is laid on the classical work because of its humanizing effect;
the lad is preparing to read ancient authors in the original, to become
acquainted with the philosophy and to partake of the refinement found
only in writers of antiquity when the influence of the shop and the love
of money were not reflected throughout literature.

The other school in bitterness of spirit speaks scoffingly of these
claims and denies that the classical languages are taught in our schools
and colleges; its advocates challenge the defenders of the older system
to produce the graduates of the ordinary college courses who can read
ancient philosophers in the original; they assert that, of college graduates who have spent from eight to ten years in the study of Greek and
Latin, only a small percentage can take a work previously unread and
read it with any degree of ease; they assert that two thirds of the college graduates are unable to read their diplomas; they refer unpleas-
antly to the statement that in theological seminaries, text-books in
scholastic Latin were abandoned not so much because the theology was
antiquated as because the students were found to be studying Latin instead of learning the theology : they prove that while the great works
of antiquity, unless in the Bohn library, are sealed books to the ordinary classical student, the works of French, German and Spanish authors are not sealed books even to those who have spent very much less
time in the study of those languages—and this too in spite of the com-

plexity of German and French idioms   They assert that Greek and Latin are taught as mere abstractions, that instead of Greek and Latin, there is taught a universal grammar, for which German or English could be used, for which Goethe or Shakespeare would answer as well as Sophocles or Horace.   They assert too that this method of training is unjust to the man; that thereby it is possible for men to enter the Christian ministry or to be admitted to the bar, even though ignorant of the simplest processes of nature and of the most commonplace facts in agriculture and the mechanical arts; that men who pass through college courses and enter upon business pursuits, show unfitness for concrete things and lose valuable time in learning to utilize their mental training.   They maintain that a study of God's works of creation is a vastly better occupation for the present and for the future than is the study of the human intellect, which, by some accounts, has fallen sadly from its first estate and by others has risen none too far above it.

As in very many other cases, the truth lies between these extremes, but it lies nearer to the modern school than to the other—a truth which has gained recognition rapidly during the last score of years, as appears from alterations in the college curriculum.   The times have changed and our methods must change with them.   Two centuries ago Latin was the common language of learned men and its place in the curriculum was as important as French and German should be now—and for the same reason.   But that reason no longer avails for the retention of Latin in its exceedingly prominent place.   Greek is necessary still for the the theologian just as is Hebrew, which is begun usually in the seminary, though a wise regard for the needs of theological students has led some colleges to place it among the electives.   The great value of Greek and Latin as now taught lies in the polish imparted; the teaching does little toward expanding the intellect, it tends rather to make the mind great in little things; its place is not at the beginning but at the close of training.   The intellect must first be shaped, then polished; the great effort prior to the college course must be to develop: true training will endeavor to assist, not to thwart nature.

In the earliest training, the studies of greatest prominence should be such as to aid the natural order of development; elementary botany, mineralogy and zoology have materials everywhere, alongside of every path.   Observing under the care of a teacher, who knows not merely what, but also how to observe, leads to the habit of comparison; the relationship of groups becomes apparent and how to make inductions

respecting cause and effect is learned—the most important of all preparatory stages; after these the study of one's self comes naturally,
first of the tangible self and then of the intangible thinking self; for
elementary psychology is as attractive to a youth as is elementary physiology, and no more difficult. In this manner, while the process of
gathering knowledge goes on, there advances with it the process of enlargement and strengthening, while the process of refining is not neglected in these and associated studies—it is only subordinated. But
a time comes when more than mere guidance, more than a gentle
effort to prevent irregular development is needed, when native tendencies show themselves too strongly and restraint or positive direction is
necessary; the process of culture, thus far merely incidental, must become prominent   And here is the place of the college.

The main object of college work is not to train men for their life
work but to prepare them for receiving such training—a fact too often
forgotten now, when colleges are endeavoring to engraft university conditions upon the college curriculum. The question is not what will be
best suited to the man's intended pursuit, but what will make him best
able to receive and profit by the immediate preparation for that
pursuit.

Yet, while recognizing this as the main object, we must not neglect
another consideration. Life at best is very short, and the portion spent
in college, from 17 to 21, is that during which, upon the whole, the
mind is most receptive, retaining, as it does, a great part of the absorb-
ing power characterizing childhood, while it has gained not a little of
the ability to acquire by reasoning. It is wrong to permit this portion
of life to pass without giving opportunity to acquire knowledge. We
live in a time when men are expected to leap into active service at
twenty-five; when opportunities for readily increasing one's stock of
general knowledge disappear quickly after life's work has fairly begun.
No wonder that we hear so often the cry of *cui bono?* respecting the
older and even respecting some of the newer modes of training   Not
a few of those who believe that language and mathematics can be
taught and should be taught so as to cultivate the very faculties reached
especially by natural science studies, are inquiring earnestly, Why
should so much of life be spent in the mere process of getting ready
to get ready? Surely something of real service beyond mere training
should be acquired during the process. The curriculum should be
prepared with this matter in view, as far as is possible, without inter-

fering with the main object of college work. We are told often that a man who spends an hour in sharpening his ax is likely to do more and better work during the day than the man who refuses to spend the morning hour in sharpening; but the man must have his breakfast as a prerequisite. In the case of the human mind, the implement and its user cannot be separated, they are one - and this is where the simile fails, despite its frequent use as an end of all argument. The man who spends all his time only in sharpening is less likely to do the full tale of work than is the one who ate a good breakfast and neglected the sharpening. But given the sharp ax and the better breakfast, there can be no doubt as to the quality of the work. Strong man and sharp ax together answer to the human mind, strong, cultured and well furnished.

Thoughtful men feel that there is a serious defect somewhere in our methods; keen, bright students find many of their studies irksome, and a few of them attractive, despite the fact that oftentimes those teaching the attractive studies are less skillful than the others. Long ago, the wise man told us that much study is weariness to the flesh; but certainly it is no more a weariness than are baseball, football, cricket, boating, foot-races or squirrel hunting; physical exercise of these types is taken with a zest which all understand. And all understand equally well that exercise thus taken is vastly more beneficial than the irksome exercise of the daily " constitutional" taken under the direction of a physician. There is no reason why mental exercise, to be beneficial, should be irksome, should have the task feature prominent. The difficulty in the curriculum lies in the undue proportion of certain types of study.

The preponderance of studies looking to culture is far too great— studies without apparent relation to present or future conditions as far as the student can see, even toward the end of his course. No matter how willing a man may be to work, he cannot work heartily if there be no apparent result; the most hopeful of men needs a little occasional fruition to keep him up; pounding a log with the blunt end of an ax is not half so cheerful work as chopping. The curriculum should commend itself, in *some* degree at least, to the intelligence of the student as of practical value, for interest is a vastly better incentive than discipline. More stress should be laid on such studies as geology, physics, chemistry and biology, including here psychology, of which now only the merest elements are taught in the arts courses of many leading col-

BALANCE ROOM

leges; such studies should not be subordinated to the so-called culture studies for they are culture studies themselves of a very high type; the study of universal grammar leads to precision, that of these studies even more; the former requires delicate perception of resemblances and differences, the latter even more; the former strengthens verbal memory, so does the latter while strengthening and cultivating the sense memory also; but the study of the latter does far more, if properly conducted; it leads the man to become careful of his positions, to be cautious in making inductions, to be less tenacious of his opinions and to be tolerant of the opinions of others, that is, to maintain a judicial frame of mind. At the same time, scientific studies should not predominate greatly in the curriculum lest irregular development result. At graduation the man ought to have laid a foundation for whatever pursuit he is to follow; he should have the polish and ease coming from the study of language and literature, the logical mode of thought coming from the study of mathematics, with the knowledge, strength and judicial tendency coming from the study of inductive sciences.

Yet this is not all that should result from college training. No mere collegiate course, though it embrace the best features of all, can lead sufficiently to such breadth of view as will enable the student to make special application of his knowledge or of his methods to every day problems. Such training is left ordinarily for post collegiate years, but it would be vastly better if some were received during collegiate years. It can be imparted by means of the so-called technical courses, say, for example, mining or civil engineering. Those courses require a very thorough knowledge of the general studies; no parrot like preparation in mathematics, mechanics, geology or chemistry will avail; the principles must be understood so as to be used readily and to be applied in all their bearings and relatioas with accuracy and despatch; for problems are presented to students which involve consideration of events apparently remote, of conditions apparently unrelated, and the reconciliation of forces apparently the most antagonistic. The whole process is that followed in later life, when a man must determine his procedure in business or in professional work by careful consideration of present conditions in the light of experience. This is recognized in France, where only a small proportion of those taking technical courses do so with the intention of making them the basis for a professional career.

By such combinations of studies can be given the training which will fit the average man for the duties of life and which cannot fail to render the feeble man much less incompetent to make his struggle for existence.

At the dedication of this beautiful building, we cannot fail to foresee some of the advantages which must accrue from the muificence of its founder.   Scientific men and the public throughout this region will be brought more closely together, a familiarity, unlikely to breed contempt on either side, but likely to lead each to learn from the other to cultivate a due humility.   It will aid in gaining a hearing for scientific men and in assigning the so-called "practical man" to his own place ; it will remove prejudices and will protect the community from great loss of money and of comfort; the place of the several departments of science will be understood and the good people of Ohio will not expect a botanist to determine the worth of a coal estate, a geologist to discover the habits of insect plagues or to discover means for their extermination, a physiologist to discuss the best localities in which to bore for natural gas, or a naval officer to make the preliminary reconnaissance for a railway route.   Before long, there will be no danger that the Legislature will be asked to investigate the honesty of a noble and devoted State Geologist because he warns the state against the sinful waste of a great blessing, such as natural gas.   In a word, the influence of this School of Science will hasten the fall of the charlatan who fattens on the ignorance and cupidity of the community.

And now, may I say a word to the students and to the alumni of Denison University ?   The ultimate, absolute success of this institution will depend largely upon you.   The tie between alumnus and Alma Mater, which some affect to ridicule, is genuine and material   Be the fees what they may, they never suffice to defray the cost of instruction; in American colleges, the cost per student is from three to even twenty times the fee, the latter proportion being that in state institutions, where fees are very small.   The indebtedness therefore is not ideal but real. Let the alumni hold this school very near to their hearts; let them make its museum, let them build up its library ; and as their prosperity increases, let them help it in other ways, that it may become stronger and stouter, able to do better work in each decade than it did in that preceding.   New friends may be raised by an energetic president, but unless the alumni form a constant stream of thoughtful care takers, the burden of chasing for new friends becomes ere long a thankless task. But if you cherish and support you Alma Mater, there will be no difficulty in crying *Esto perpetua* with a sturdy faith that the prayer will be answered.

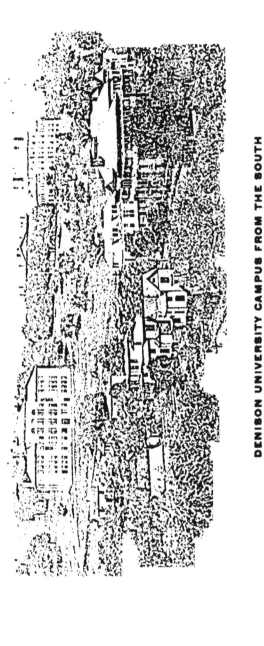

DENISON UNIVERSITY CAMPUS FROM THE SOUTH

BARNEY MEMORIAL SCIENCE HALL.

Barney Memorial Hall is located in a beautiful situation, south of the library building and the Academy dormitories, on the elevated slope above the buildings of Shepardson College. It is built of the best Zanesville buff pressed brick, with the lower story and trimmings of buff Amherst stone. It consists of a main central portion 50x70 feet, four stories high, with two wings, each 45x36 feet and three stories high The construction is very massive, with two thick interior walls running lengthwise through the main portion and two others separating the main building from the wings. The building is thus divided vertically by heavy brick walls into five distinct parts. The foundation rests for the most part on the bed rock below. This solidity of construction is important in securing the necessary freedom from vibration for the use of delicate instruments.

The building is heated by steam, partly by direct radiation and partly by indirect. A high pressure boiler is used, with an automatic reducing valve which keeps the pressure on the building at any desired point. Good ventilation is secured in the following way: in the first place by introducing fresh air at six different points to steam coils from which it passes to the rooms above ; in the second place by a system of foul-air registers which communicate with two brick stacks about 75 feet high, surrounding the high cast iron chimneys connected with the boiler. The heat of the inner iron chimneys produces a strong draft in the outer stacks which rapidly removes vitiated air from the rooms.

The building was designed by Peters & Burns, of Dayton, in the colonial style of architecture, to suit the special requirements of laboratory use. It contains forty rooms, most of them adapted to some distinct purpose. They are plentifully provided with blackboards, sliding chart frames, dust tight cases, gas, water, steam, slate-topped tables and fume cupboards for carrying corrosive or disagreeable gases from the working rooms. The plumbing is very complete, water being distributed to about sixty points in the building and gas to many more than than that. All sinks are provided with traps that can be readily opened and the plumbing of the toilet rooms has been carried out in the most approved manner.

The general electrical equipment of the building occupies a large room in the basement and consists of a 55 horse-power boiler, a 50 H. P. Ball engine, direct-connected to a 40 K. W. Thresher multipolar dynamo, which gives a current of 350 amperes at 115 volts This plant is in operation every evening, supplying light to a number of dormitory buildings and at the same time, with the aid of a 10 H. P. " booster " dynamo, charging a large storage battery which consists of 60 cells of chloride accumulator of 600 lamp-hour capacity. This battery supplies current for light during the remainder of the night, and the next day for general laboratory uses, including current, lights, power for the shop motors, and heat for special purposes.

The gas used in the building is gasolene gas and is supplied to all the rooms by a Detroit gas machine of 75-light capacity. It is used principally for heating, in Bunsen burners, gas stoves and assay furnaces for testing ores ; but it also supplies light to some rooms not yet wired for electricity.

Three electric lanterns and two complete outfits for producing the lime light, give good opportunity for illustrating all subjects by stereopticon projections. The numerous appliancs peculiar to the several laboratories will be described in connection with the special accounts of those laboratories.

Besides the first cost of about $40,000 for construction, over $15,-000 has been spent for equipment.

DEPARTMENTS OF PHYSICS AND CHEMISTRY.

THE DEPARTMENT OF PHYSICS occupies ten rooms, most of which are on the south side of the main building. Its equipment has cost about $7,000. The lectures in general physics are given in a large room (45x26 feet) on the first floor. This is a laboratory and lecture room combined (marked " General Physics " on the plan). A large apparatus case, 20x5x8 feet, enclosed by glass doors so as to be accessible from either side, almost bisects the room near the center. The east end of the room is used for lectures and recitations, the other end for individual laboratory work. In this way apparatus is placed so as to be convenient for either purpose. The lecture portion contains seats for 35 students, a long demonstration desk, furnished with a tank-sink, gas and wires from both the dynamo room and laboratory room. An electric lan-

GENERAL PHYSICAL LABORATORY

tern always stands ready for projections, and a porte-lumiere in a south window enables one to throw sunlight wherever it may be desired. A beam stretches across the room above for heavy suspensions. The other end of the room is provided with laboratory desks. One along the south side of the room is 35 feet long. A wider table runs nearly across the room at right angles to the first Beside this there is a large stone slab supported upon a heavy brick pier, which passes through an opening in the floor to the solid rock 17 feet below. This gives a support for delicate instruments that is free from floor-vibrations. From one corner of this large room a photometry room, 12x5, is cut off. This can be made perfectly dark, for measuring the intensity of various sources of light and other optical work. Another small room, 14x8 feet, is connected by an arched space to the large room and can be readily cut off for special work. It contains a stone table set in a corner made by two 24-inch walls for steadiness, a sink and a large fume-cupboard with a tile conduit to the draught-stack, also one or two moveable tables.

On the second floor are a number of rooms for more advanced work. One (marked "Advanced Physics") 26x17 feet, is furnished with a sink, about 15 gas terminals, electric wires from the dynamo room, a table across one end supported by heavy brackets from the outer brick wall, and a number of very solid moveable tables. A suspension beam runs above through the length of the room. This room contains much of the finer apparantus, as the dividing engine, standard meter, various certified standards of electrical resistance, potential, and capacity, standard thermometer, heliostat, reflecting galvonometers, mercury pumps, induction coils, X-ray outfit, saccharimeter, precision balance, etc. A research room opening from it is equipped with water, gas, electricity, stone shelf set in brick walls and a special low-voltage circuit from the battery room. The room marked "Study" (13x11 feet) contains also a small department library for advanced students. Opposite this is the photographic dark-room of this department. It contains two sinks, gas, electricity and light tight closet.

In the basement beside the engine and dynamo rooms, there are three rooms, completing the physical equipment. One is a large shop, 22x21 feet, for apparatus construction. It is provided with a 3 H. P. electric motor which takes current from the dynamo-room for running the line of shafting, which distributes power to two lathes (one screw-cutting), a circular saw, emery wheel and polishing head. This shop is well stocked with hand tools for both wood and metal work, and a

forge stands in an adjoining room. The next room is provided for certain kinds of electrical testing connected with the study of dynamo-
electric machines. It is not yet fully equipped. Near the other end
of the basement is a small room about 9x7 feet, surrounded by heavy
stone walls and projecting back into the earth. This construction makes
its temperature very constant and it is to be used as an even temperature vault for carrying on work which needs to be done at a constant
temperature.

THE DEPARTMENT OF CHEMISTRY occupies the west wing
of the building. It contains eight rooms, none of them directly connected with rooms of any other department, so that the odors peculiar
to a chemical laboratory do not give trouble elsewhere. The lecture
room and beginners' laboratory is a combination room on the first floor.
It is 36x33 feet in size, provided at one end with a long demonstration
desk, fume-hood, apparatus cases, sliding chart frame and blackboards,
in the centre are seats with writing arms for about forty students, and on
the remaining three sides 24 working desks, each with water, gas and
chemicals, also two more hoods and reagent cases. Electric wires bring
current at low potential (six volts) for electrolytic work, from the battery room below. The analytical chemistry occupies a room on the
second floor, 36x18 feet. It has a demonstration desk, working desks
equipped with water, gas and reagents for 24 students, fume hoods,
drying ovens, steam coils, apparatus for distilled water, etc. Adjoining this is the organic laboratory, 20x14 feet, equipped with desks,
sinks, hood, etc., especially arranged for organic synthesis. It is used
also by advanced classes in water analysis, gas analysis, electro chemistry, etc A fine set of Hempel apparatus for gas analysis has been added
recently, also Beckman's apparatus for determining molecular weights
by the lowering of the boiling point of solutions. Both these rooms have
permanent connections for 6-volt current from the battery-room and an
auxiliary storage battery is kept in this room. Beyond is the balance
room, 15x19, containing four analytical balances and several for other
purposes. A chemical stockroom, 15x5 feet, opening into both the
analytical and organic rooms, is well furnished with chemicals. It can
be shut up perfectly dark, and is therefore used by qualitative students
for work with the chemical spectroscope. For general supply there is
a larger stockroom, 21x10 feet, in the basement. Adjoining this is a
fire-proof acid vault, 12x8 feet, for storing large quantities of acids,

GENERAL CHEMICAL LABORATORY

corrosive or highly inflammable substances. The assay-room, 14x14 feet, is also in the basement. It contains a long desk on one side, shelving for chemicals, a coke muffle-furnace, gas muffle and crucible furnaces, a good supply of tools, scorifiers, cupels, etc. An electric furnace for highly refractory substances, has just been placed in this laboratory.

DEPARTMENT OF GEOLOGY AND BOTANY.

THE DEPARTMENT OF GEOLOGY occupies three large laboratories exclusive of the museum.

The general laboratory of geology (18x24 feet) is in the southeast corner of the east wing on the basement floor. It is equipped with case for illustrative material, chart and map cases, charts, maps, globes, models, drawing tables with elevating tops and other necessary apparatus. The laboratory of mineralogy (18x26 feet) is also in the basement and on the south side of the main building. It is furnished with slate topped work tables, with tin lined drawers and reagent racks in middle of the tables, gas, blowpipes, blast lamps, fume hoods, micro scopes and a good collection of working minerals. This laboratory is also used for the microscopic work in lithology; lithological microscopes and a good library of slides of igneous and sedimentary rocks are at the disposal of students. The lithological lathes and saws are in the special shop of the geological and biological departments in an adjacent room.

THE DEPARTMENT OF BOTANY occupies two rooms on the north side of the main portion of the building on the first floor. The laboratory of phenogamic botany (27x27 feet) is planned to accommodate 24 students. The portion of the room next to the large north windows is furnished with convenient microscope desks which give a cupboard and two drawers to each student, who is also provided with a compound microscope of modern pattern with full set of objectives and eye pieces. The back portion of the room is seated with chairs and serves as a lecture room. A herbarium case occupies the wall space on the south side. A fume cupboard with slate table top fills the corner next to the ventilating stack and connects with it. A water sink occupies the opposite corner. A large black board runs the entire length of the west wall and above this is a sliding chart rack similar to all the other racks in the building. There is also a large chart case provided with a goodly number of excellent botanical charts. The room is amply provided with water, gas and electricity. The laboratory of cryptogamic botany (19x17 feet) is adjoining. The microscope desks are arranged next to the walls under the windows and are of the same plan as those in the larger room.

The laboratory is fully equipped with the best compound microscopes, microtomes, sterilizers, incubators and a large library of microscopic slides. A large herbarium case contains the cryptogamic herbarium.

The biological photographic dark room and the stock room both open off from the cryptogamic laboratory. A full set a photographic apparatus and a well equipped dark room provided with two sinks, washing tanks, etc., and nicely illuminated with electricity and gas, with sliding colored glass windows giving various colored illumination, affords good opportunity for experimental photographic and microphotographic work.

This dark room is used also by the engraving department for their photographic work.

DEPARTMENT OF ZOOLOGY.

THE GENERAL ZOOLOGICAL LABORATORY (26x27 feet) is a well lighted room on the second floor with north and east exposure. The work tables are arranged along the two outer walls, the remainder of the floor space being used as lecture room. The room accommodates 21 students and each desk is supplied with individual lockers, compound microscope and the conveniences for dissection. Charts, models and mounted skeletons, human and comparative, are supplied, and the wall cases contain those specimens which are most useful in demonstrating zoological types. Much of this material is, however, stored away, pending the time when suitable cases can be supplied for its reception in the museum. A small reference library is also provided.

THE ADVANCED ZOOLOGICAL LABORATORY (17x26 feet) adjoins the General Laboratory and is lighted from the north and west. The work tables are bracketed to the outer brick walls to insure the steadiness necessary for high-power microscopic work. The central floor-space is occupied by a large slate-topped table fitted up with the baths and reagents necessary for the embedding the sectioning of tissues for the microscope. Good microscopes of modern pattern, immersion lenses, and several of the most approved types of microtomes, together

ZOOLOGICAL LABORATORY.

with a full list of preserving and staining reagents are supplied. An incubator, sink and slide cabinet complete the furnishing The latter contains a large collection of microscopical preparations illustrating the tissues of the various groups of vertebrate and invertebrate animals and is especially complete in the vertebrate nervous organs.

THE LABORATORY OF PHYSIOLOGICAL AND COMPARATIVE PSYCHOLOGY (26x27 feet) is a large room on the third floor. It contains a Ludwig kymograph, chronoscope, pendulum myograph, time markers, with proper electrical connections, and numerous other pieces of apparatus, many of which were constructed in the machine shop of the department. The equipment is, however, as yet very incomplete. The courses thus far given have been out-growths of the neurological laboratory and in the furnishing of this laboratory attention is directed rather to the requirements of the physiological and comparative aspects of the science than to those of experimental psychology in the wider sense. The library is abundantly supplied with the current neurological and psychological literature and through the medium of the Journal of Comparative Neurology the department is kept in touch with the most advanced movements in these departments.

THE INJECTORIUM (9x12 feet). This is a small room is the basement with cement floor and walls designed as a preparation room for the department of Zoology. In it all rough dissection, injection of specimens etc., is done and alcoholic specimens are prepared and dissecting material stored. It is supplied with the gravity injecting apparatus designed by Professor Tight, sink, work table, tanks and cases.

THE SPECIAL SHOP of the departments of Zoology and Geology occupies a room (17x26 feet) in the main portion of the building on the south side, on the basement floor. It is supplied with an electric motor of 2 horse power which drives the main shaft, which latter is supported by hangers resting on the floor. Besides a good assortment of wood and metal working tools, benches, etc , the power machines include a Royle combination saw, a large screw cutting metal lathe, a wood lathe, a Royle former, a lithological lathe of the Royle pattern and a Royle router. This shop is used by the departments of zoology and geology for the construction of apparatus and by the engraving department for the routing and blocking of plates.

GENERAL ROOMS.

THE MUSEUM room occupies the entire East wing on the first and second floors. The second floor consists of a broad gallery around the entire room, thus giving the central space on the first floor two stories high to the ceiling.

The intention is to devote the first floor largely to geology and paleontology and the second floor gallery to zoology. The collections are not at present in place owing to the lack of funds to supply the necessary cases. A large amount of material is on hand but is for the most part stored away to protect against loss and injury until the cases and furnishings for the museum can be secured.

A RECITATION AND LECTURE ROOM (17x26 feet) furnished with desk and writing arm chairs, black boards, chart racks, etc., is utilized by the various departments.

THE DENISON SCIENTIFIC ASSOCIATION ROOM (26x46 feet) occupies the entire south side of the main building on the third floor. The Association holds its bi-weekly meetings here. The room is well seated and supplied with black boards, projection screen, demonstrating desks and sink, electricity and gas, thus making it a very convenient place for the presentation of papers needing illustration or demonstration. At one corner of the room the lift shaft communicates with each floor. The room is also utilized for lecture work by the different departments.

THE SCIENTIFIC LIBRARY ROOM (18x26 feet) is on the third floor and contains the exchange library of the Bulletin and Journal of Neurology, consisting almost entirely of technical scientific literature.

THE ATTICS are two in number (46x36 feet each) and are easily accessible from the third floor main hall, as they are over the two wings. They furnish very convenient storage rooms.

THE LAVATORIES are on the second floor and in the basement and are well furnished.

THE BOILER AND FUEL ROOMS are in the north side of the main building next the area way. They are entirely enclosed from the rest of the building by thick brick walls and tile ceilings except through the door leading through the engine and dynamo room, thus furnishing good protection against fire. Chemical fire extinguishers are also placed in the halls on every floor.

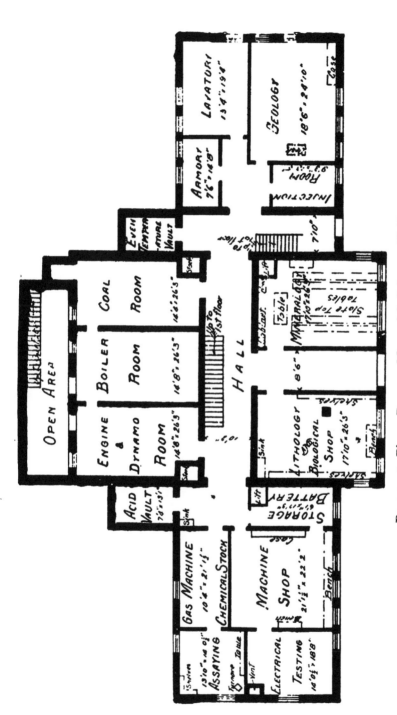

Basement Plan, Barney Memorial Science Hall.

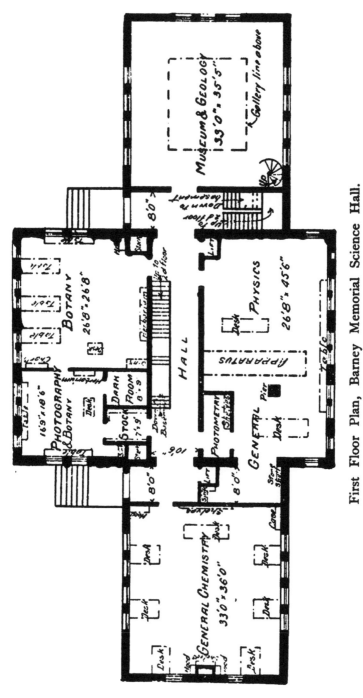

First Floor Plan, Barney Memorial Science Hall.

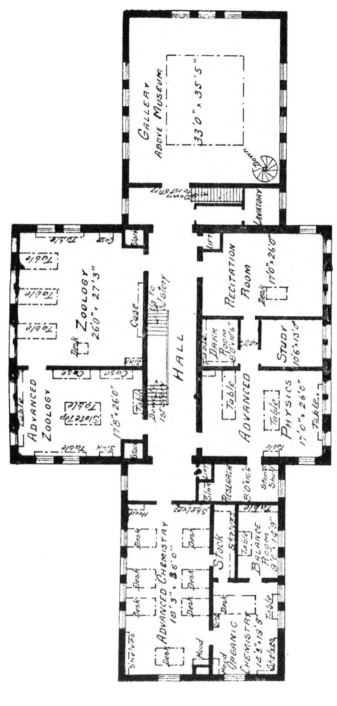

Second Floor Plan, Barney Memorial Science Hall.

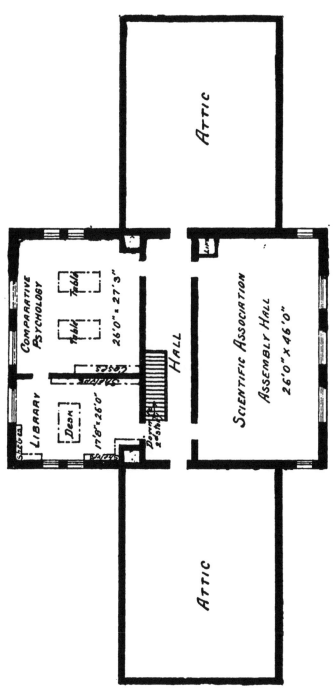

Third Floor Plan, Barney Memorial Science Hall.

BULLETIN

OF THE

SCIENTIFIC LABORATORIES

OF

DENISON UNIVERSITY.

EDITED BY

THOMAS L. WATSON,

Permanent Secretary Denison Scientific Association.

GENERAL INDEX TO THE FIRST TEN VOLUMES OF THE BULLE-
TIN OF THE SCIENTIFIC LABORATORIES OF DENISON
UNIVERSITY. FROM 1885 TO 1897 INCLUSIVE.

By W. W. STOCKBERGER

Stockberger, W. W.
General Index to the First Ten Volumes of the Bulletin of the Scientific Laboratories of Denison University. From 1885 to 1897 inclusive. Bulletin Scientific Laboratories of Denison University, Granville, Ohio, August, 1904, pp. 1-39.

General Index to the First Ten Volumes of the Bulletin Scientific Laboratories of Denison University. From 1885 to 1897 inclusive. By W. W. Stockberger. Bulletin Scientific Laboratories of Denison University, Granville, Ohio, August, 1904, pp. 1-39.

Bulletin Scientific Laboratories of Denison University, General Index to the First Ten Volumes of the. From 1885 to 1897 inclusive. By W. W. Stockberger. Bulletin Scientific Laboratories of Denison University, Granville, Ohio, August, 1904, pp. 1-39.

Denison University, General Index to the First Ten Volumes of the Bulletin of the Scientific Laboratories. From 1885 to 1897 inclusive. By W. W. Stockberger. Bulletin Scientific Laboratories of Denison University, Granville, Ohio, August, 1904, pp. 1-39.

GENERAL INDEX

TO THE

FIRST TEN VOLUMES

OF THE

BULLETIN

OF THE

Scientific Laboratories

OF

Denison University

From 1885 to 1897, inclusive.

———

BY W. W. STOCKBERGER.

All references are brought under one alphabet. Names of new species are followed by n. sp. Italicised page number indicates a reference to an illustration. The Roman characters I and II following the volume numerals 8 and 9 refer to the parts in which these volumes were issued, a distinction not observed in the other volumes since the pagination of each is consecutive. Figures not referred to in the text are indexed by number of volume and plate. The letter T preceding a numeral refers to the tables in Vol I.

NOTE: Plates VIII, XV, and XVI of Vol. II were issued as the last three plates in Vol III.

Plate XI, wanting in Vol. III, appears as plate of same number in Vol. IV.

INDEX.

A

(5)

nodocarinatus *2:18*.
percarinatus *2:17*.
perelegans *3:90*.
pulchellus **2**:19.
sp. *3:90*.
stramineus **2**:19.
sub–cordiformis n. sp. *2:18*.
Bellflower **7**:56.
Bent grass **7**:95.
Beomyces roseus **9**–I :14.
Berberidaceae **7**:15.
Berberis vulgaris **7**:15.
Berea grit **4**:107.
    shale of Ohio **4**:107.
    shale exposed at Moot's
        Run, O. **5**:26.
Bergamont. Wild **7**:69.
Bidens bipinnata **7**:52.
    cernua **7**:52.
    connata **7**:51.
    chrysanthemoides **7**:52.
    frondosa **7**:51.
Bignoniaceae **7**:66.
Big Sandy Valley. Preglacial drain-
    age in **9**–II :26.
Bindweed. Black **7**:74.
Biological notes on Fiber, Geomys
    and Erethizon **6**:15.
Biotite **1**:T8.
Bishop's Leaf **7**:36.
    Cap **7**:36.
Bittersweet **7**:63.
Bittersweet. Climbing **7**:26.
Blackberry. High **7**:33.
    Low **7**:33.
Black Gum **7**:43.
Black Hand. Conglomerate of **9**–I :9.
Black Hand rock **8**–II :40.
Black Haw **7**:44.
    Medick **7**:29.
    Snakeroot **7**:13.
    Sugar Maple **7**:27.
Bladder cells of the trigeminis of
    Arctomys **5**:61.
Bladder Ketmia **7**:24.
Bladder-nut. American **7**:28.
Bladderwort **7**:66.
Blazing Star **7**:86.
Blephilia ciliata **7**:69.
    hirsuta **7**:69.
Blood Root **7**:16.
Blue Bells **7**:61.
Blueberry. Low **7**:56.
    Swamp **7**:56.
Bluebottle **7**:56.
Blue clay of the Clinton **1**:68.
Blue Cohosh **7**:15.
Blue-eyed Grass **7**:83.
Blue-eyed Mary **7**:64.
Blue Flag **7**:83.

Blue Grass **7**:97.
Blue Weed **7**:62.
Boehmeria cylindrica **7**:77.
Boneset **7**:46.
Borraginaceae **7**:61.
Bosmina cornuta **8**–I :4.
    atlantaensis n. sp. **8**–II :23.
Botanical laboratory. Notes from **8**–
    II :7.
Botany. Dept. of at Denison Univ.
    10:87.
Botrychium ternatum var. interme-
    dium **7**:101.
    virginianum **7**:101.
Boughton, W. H. **10**:15 — Biog. *36*.
Boulder clay near Beech Flats, O.
    9–I :28.
Bourneville, O. Preglacial channel
    near **9**–I :19.
Box Elder **7** :28.
Brachionus bakeri **1** :55, *6:64*.
    intermedius **1** :56.
    militaris *1:56*, **6**:65.
    pala **6** :63.
    tuburculus n. sp. *6:65*.
    urceolaris **6** :64.
Brachiopoda **2** :105.
    in Clinton group of Ohio **1**:
        78.
    of Flint Ridge, Ohio **2** :43.
    of Licking Co., Ohio. Key to
        4 :11.
Brachyelytrum aristatum **7** :95.
Brachymetopus **2** :53.
    discors **2** :57.
    hibernicus **2** :55-57.
    lodiensis *2:55*.
    mc coyi *2:55*.
    ouralicus *2:55*.
Brain of Arctomys, external form and
    measurements of **5** :51.
Brain. Comparative structure of
    Arctomys and Didelphys
    5 :76.
Brain of Erethizon **6** :26.
    of Geomvs **6** :26.
    method of hardening **6** :27.
    of Opossum **6** :75.
    of rabbit. Summary of Stre-
        da's work on **5** :41.
    of rat. Summary of Streda's
        work on **5** :41.
    Rodent. General description
        of **5** :40.
    Studies in topography of ro-
        dent **6** :26.
Brake. Common **7** :99.
Brassica nigra **1** :31, **7** :19.
    sinapistrum **7** :19.
Bremen, O., drainage near **9**–II :35.

Bromus ciliatus 7 :98.
" var. purgans 7 :98.
mollis 7 :98.
secalinus 7 :98.
tectorum 7 :98.
Bronzite 1 :T10.
Brooklime. American 7 :65.
Brookweed 7 :57.
Brown's Quarry, Clinton group of 3 :10.
Brunella vulgaris 7 :70.
Bryozoa of Flint Ridge, O. 2 :71.
of Clinton group of O. 2 :149.
of Waverly group of O. 4 :63.
Bucania exigua n. sp. *1 :99*.
trilobata *2 :103*.
Buckeye, Ohio 7 :27.
Buckwheat, 7 :74.
Buds. Superposed 1 :25.
Buellia parasema 9–I :14.
petraea 9–I :14.

Bugle Weed 7 :68.
Bulletin Scientific Laboratories of Denison University. Contents 10 :41.
Editor 10 :19.
Exchange list 5 :4, 10 :41.
Founding of 10 :16.
Bullrush 7 :91.
Bunch–berry 7 :43.
Bur–cucumber 7 :39.
Burdock 7 :53.
Burlington group 4 :99.
Bur Oak 7 :78.
Bur Marigold 7 :52.
Burning–bush 7 :26.
Buttercups 7 :13.
Butter-and-eggs 7 :64.
Butterfly–weed 7 :58.
Butter-nut 7 :77.
Butter–weed 7 :53.
Bytownite 1 :T18.

C

Cacalia atriplicifolia 7 :53.
reniformis 7 :53.
Calanidae 6 :69.
Calcedony 1 :T19.
Calcite 1 :T7.
Calcium. Crystal system of 1 :130.
Fluo–silicate of 1 :130.
California Valley, O. Preglacial drainage in 9–II :27.
Callopora magnopora n. sp. *2 :173*.
ohioensis n. sp. *2 :174*.
punctata 4 :91. *See* Leioclema punctatum.
Callosum of Geomys and Erethizon 6 :38.
of Opossum 6 :81.
Calopogon pulchellus 7 :81.
Caltha palustris 7 :13.
Calymene 1 :109.
——— *1 :109*.
blumenbachii ? *1 :110*. *See* C. vogdesi.
clintoni *see* C. vogdesi.
niagaraensis 1 :109.
vogdesi n. sp. *2 :95*.
Camassia fraseri 7 :84.
Cambrian formation 9–I :4.
Camp Corwin, O. Clinton exposures of 3 :11.
Campanulaceae 7 :55.
Campanula americana 7 :56.
aparanoides 7 :56, 8–II :4.
Camptosorus rhizophyllus 7 :100.
Camtocercus macrurus 8–I :4.
Canada thistle 7 :54.
Canalis centralis of Arctomys 5 :56.

Canary–grass 7 :95.
Cancer-root 7 :66.
Cancrinite 1 :T7.
Cannabis sativa 7 :76.
Candona acuminator 8–II :19.
crogmaniana n. sp. 8–II :20.
delawarensis n. sp. 8–II :21.
Canthocamptus 1 :37.
Caprifoliaceae 7 :43.
Capsella bursa–pastoris 5 :10, 7 :19.
Caraway 7 :41.
Carum carui 7 :41.
Carboniferous trilobite 2 :51.
Cardamine hirsuta 7 :17.
rhomboidea 7 :17.
" var. purpurea 7 :17.
Cardinal flower 7 :55.
Cardiopsis ovata *4 :38*, Fig. 5 *not* Fig. 6.
Carex bromoides 7 :93.
cephalophora 7 :93.
divisii 7 :92.
echinata var. cephalantha 8–II :5.
gracillima 7 :92.
granularis 7 :92.
grayii 7 :91.
hystricina 7 :92.
interior var. capillaceae 8–II :5.
intumescens 7 :91.
laxiflora 7 :92.
" var. patulifolia 7 :92.

laxifolia var. styloflexa 7:93,
    8-II:5.
    "   var. varians 7:92.
lupulina 7:91.
    "   var. pedunculata
    8-II:5.
lurida 7:92.
oligocarpa 7:92.
pennsylvanica 7:93.
plantaginea 7:93.
platyphylla 7:93.
prasina 7:92.
pseudo-cyperus 7:92.
    "   var. ameri-
    cana 7:92.
rosea 7:93.
shortiana 7:92.
sparganioides 7:93.
squarrosa 7:92.
stenolepis 7:92.
stipata 7:93.
tribuloides 7:93.
trichocarpa 7:92.
utriculata 7:92.
varia 7:93.
virescens 7:92.
vulpinoides 7:93.
Carpet weed 7:39.
Carpinus caroliniana 7:78.
Carrion flower 7:84.
Carya alba 1:29, 7:77.
    amare 1:30.
    microcarpa 1:30.
    olivaeformis 1:30.
    porcina 1:30, 7:78.
    sulcata 1:30, 7:78.
    tomentosa 1:30.
Caryophyllaceae 7:39.
Cassia chamaecrista 1:32, 7:31.
    marilandica 7:31.
    nictitans 7:31.
Castanea sativa var. americana 7:79.
Catalpa bignonioides 7:67.
Catalogue of the Phanerogams and
    Ferns of Licking Co., O.
    7:1.
Catch fly 7:21.
Catgut 7:29.
Cathypna leontina n. sp. 6:61.
    ohioensis 6:61. See Distyla
    ohioensis 1:54.
Cathophyllum australl n. sp. 3:128.
Catnip 7:69.
Catskill region 9-I:5.
Cat-tail 7:88.
Caulophyllum thalictroides 7:15.
Ceanothus americanus 7:26, 8-II:3.
Cedar. Red 7:80.
Celandine 7:16.
Celastraceae 7:26.

Celastrus scandens 7:26.
Cell arrangement. Instrument for
    rapidly changing 5:16.
Cell arranger 8-II:29.
Celtis occidentalis 7:76.
Cenchrus tribuloides 7:94.
Centaurea cyanus 7:54.
Central nervous system of Arctomys
    5:35.
    of rodents 5:35.
Centronella julia 3:49.
Cephalopoda of Flint Ridge, O. 2:17.
Cephalanthus occidentalis 7:45.
Ceramopora expansa 2:169.
Chrysosplenium americanum 7:36.
Chydorus sphaericus 6:69, 8-I:6,
    8-II:25.
Cichorium intybus 8-II:4.
Cicuta bulbifera 8:41.
    maculata 7:41.
Cimicifuga racemosa 7:13.
Cincinnati group 1:67-69.
    geanticline 9-I:4.
Cinereum of Erethizon 6:29.
Cinna arundianaceae 7:96.
Cinnamon Fern 7:101.
Circaea lutetiana 7:39.
Cladocera 1:37-39.
Cladocera, Copepoda, Ostracoda and
    Rotifera of Cin., O.
    Notes on 6:57.
Cladocera of Cin., O. Notes on
    8-I:3.
    of Georgia. Notes on
    8-II:22.
Cladonia caespiticia 9-I:14.
    cristatella 9-I:14.
    delicata 9-I:14.
    fimbriata 9-I:14.
    furcata var. crispata 9-I:14.
    gracilis var. verticillata
    9-I:14.
    mitrula 9-I:13.
    pyxidata 9-I:14.
    rangiferma var. alpestris
    9-I:14.
    ravenelii 9-I:14.
    squamosa 9-I:14.
    symplycarpa 9-I:11, 9-I:13.
    uncialis 9-I:14.
Clathropora clintonensis 2:154.
    frondosa 2:154.
Clava of Erethizon 6:30.
Clay. Boulder 9-I:28.
Claytonia virginica 5:10, 7:22.
Cleavers 7:45.
Clematis viorna 7:11.
    virginiana 7:11.
Cleveland shale 4:110.
Clinochlore 1:T12.

Cyclora 1:96.
  alta n. sp. *1:96.*
Cymbellae gastroides 3:115.
Cynodon dactylon 7:96.
Cynoglossum officinale 7:61.
  virginicum 7:61.
Cynthiana, O. Glacial drift near
  9–I:26.
Cyperaceae 7:90.
Cyperus aristatus 7:90.
  diandrus 7:90.
  esculentus 7:90.
  flavescens 7:90.
  strigosus 7:90.
Cypria exculpta 8–II:13.
  inequivalva n. sp. 8–I:6,
  II:14.
Cypricardina 2:35.
  (?) carbonaria *2:35.*
Cypricardinia scitula n. sp. *4:38.*
Cypricardites 1:93.
  ferrugineum 1:93.
Cypridopsis vidua 6:73, 8–II:19.
Cypripedium acaule 7:83.
  parviflorum 7:82.
  pubescens 7:82.
  spectabile 7:83.
Cypris burlingtonensis n. sp. 8–II:17.
  crenata n. sp. 8–I:9.

fuscata 8–II:16.
herricki n. sp. *6:71.*
  " late larval history
  of 8–II:11.
incongruens 8–I:8.
laevis, *see* Cyclocypris laevis.
ovum, " " "
  sp. (?) 6:71.
striolata, *see* Cypria exculpta.
virens 6:71.
Cyrtia cuspidata, *see* Syringothyris
  cuspidatus 3:41.
  simplex, *see* Syringothyris
  cuspidatus 3:41.
Cyrtina acutirostis 4:Pl. 11.
  sp. *3:47.*
Cystodictya angustata n. sp. *4:82.*
  carbonaria *2:74.*
  lineata 2:75.
  occellata 2:75.
  simulans n. sp. *4:81.*
  sp. undet 4:83.
  zigzag n. sp. *4:81.*
Cystodictyonidae 2:165.
Cystopteris bulbifera 7:100, 8–II:5.
  fragilis 7:100.
Cythere ohioensis n. sp. *4:60.*

D

Dactylis glomerata 7:97.
Daisy, Ox–eye 7:52.
Dalmanites (?) 2:53.
  cuyahogae *2:53.*
  emmrich 1:116.
  verrucosus 1:101.
  vigilans 1:101.
  werthneri 1:68, 76, 101, 116,
  2:101.
Dandelion 7:54.
Danthonia spicata 7:96.
Daphnia 1:21.
  pulex 6:67, 8–I:4.
Darnel 7:98.
Datura stramonium 7:63.
  tatula 7:63.
Daucus carota 7:40.
Day-lily 7:84.
Dayton, O., Quarries of 1:66.
  Rock exposure near 1:66–68.
Decodon verticillatus 7:38.
Deerberry 7:56.
Delessite 1:T19.
Delphinium 1:27.
  consolida 1:32.
  exaltatum 7:13.
  tricorne 7:13.

Deming, J. L. List of diatoms from
  Granville, O. 3:114.
Denison Scientific Association 10:37.
Denison University, Curriculum of
  10:20.
  Department of Physics and
  Chemistry 10:84.
  Department of Geology and
  Botany 10:87.
  Department of Z o o l o g y
  10:88.
  Historical sketch 10:7.
  Museum 10:90.
  Scientific faculty *10:29.*
  " publications 10:41.
Dentalium sp. *2:146.*
  granvillensis n. sp. 3:92.
Dentaria diphylla 7:16.
  laciniata 7:17.
Desmodium acuminatum 7:30.
  canadense 7:30.
  canescens 7:30.
  dillenii 7:30.
  laevigatum 7:30.
  nudiflorum 7:30.
  paniculatum 7:30.
  pauciflorum 8–II:4.

E

additions and corrections
5:33.
Fowke, Gerard. Preglacial drainage
channels in Ross Co., O.
9–I:15.
Foxglove 7:65.
Foxtail 7:94.
Fragaria vesca 7:34.
virginiana 7:33.
" var. illinoensis
7:34.
Fragilaria lanceolata 3:115.
Frasera carolinensis 7:59.

Fraxinus americana 1:*27*, 7:58.
pubescens 7:58.
quadrangulata 7:58.
sambucifolia 1:*29*, 7:58.
viridis 1:*29*.
Friceratium jensenianum 3:115.
solenoceros 3:115.
venosum 3:115.
Fringillidae 1:7.
Fultonham, O., section at 3:21.
Fumariaceae 3:21. .
Fusulina cylindrica 2:15, *2:50*.

G

Galium aparine 7:45.
asprellum 7:46.
circaezans 7:45.
concinnum 7:46.
lanceolatum 7:45.
latifolium 7:45.
pilosum 7:45.
trifidum 7:45.
" var. latifolium 7:45.
triflorum 7:46.
Garlic, wild 7:84.
Garnet 1:T2.
Gas, in Lower Silurian 1:67.
Gasteropoda, in Clinton Group of O.
1:94.
of Flint Ridge 2:17.
Gaultheria procumbens 7:56.
Gaura biennis 7:39.
Gaylussacia resinosa 7:56.
Geanticline Cin. O. 9–I:4.
Gentianaceae 7:59.
Gentian, Closed 7:59.
Gentiana andrewsii 7:59.
crinita 7:59.
Geological aphorisms 4:98.
section, Ashland Co., O.
4:102.
" Lyons Falls 4:101.
" Rushville, O. 4:102
" Sciotoville, Ohio.
4:102.
Geology, chemical 3:3.
Geology, Department of, Denison
University 10:87.
Licking Co., O. 2:5-144,
3:13, 4:11, 4:97.
and Lithology of Michipi-
coten Bay 2:119.
Geomys, Fiber and Erethizon, Bio-
logical notes on 6:15.
bursarius, Biological notes
on 6:18.
brain of 6:31.

food of 6:22.
habits of 6:20.
relationships of 6:20.
Georgia, notes on Cladocera of
8–II:22.
Geraniaceae 7:24.
Geranium maculatum 7:24.
Gerardia flava 7:65.
purpurea 1:31.
Germander, American 7:67.
Germination of Phoenix Dactylifera
7:8.
Gervilla (?) ohioense n. sp. *2:36*,
2:145.
Geum album 7:33.
strictum 7:33.
vernum 7:33.
virginianum 7:33.
Gill–over–the–ground 7:69.
Gilpatrick, J. L. Biography of *10:28*.
Ginger, Wild, 7:74.
Ginseng, 7:42.
Glacial action, effect of 1:71.
epoch 9–I:2.
ice sheet, crushing effect of
6:12.
Glauconome whitii n. sp. *2:78*.
Glauconome whitii n. sp. *2:78*.
Gleditschia triacanthos 1:33, 7:31.
Glyceria acutiflora 7:98.
canadensis 7:97.
elongata 7:97.
fluitans 7:98.
nervata 7:97.
pallida 7:97.
Glyptopora megastoma 4:83.
Gnaphalium purpurem 7:49.
uliginosum 7:49.
Gneiss, near Pacolet Mills, S. C. 4:7.
of Dog River. 2:128.
Goat's-beard 7:32.
–rue 7:29.

H

tuberosus **7**:51.
Heliopsis laevis **7**:50.
Helminthite **1**:T12.
Hematite **1**:T8.
Hemerocallis fulva **7**:84.
Hemipronites **2**:12.
crassus *2:50.*
crenistria *3:37, 4:24.*
Hemitrypa ulrich n. sp. *2:152.*
Hemlock **7**:80.
Hemp **7**:76.
Hepatica acutiloba **7**:12.
triloba **7**:11.
Heracleum lanatum **7**:40.
Hercynite **1**:T3.
Herrick, C. Judson **10**:19.
Biography of *10:34.*
(and C. L. Herrick) Biological notes upon Fiber, Geomys and Erethizon **6**:15.
Editor Journ. Comp. Neurol. **10**:54.
Studies in the Topography of Rodent Brain **6**:26.
Writings of **10**:34.
Herrick, Clarence L. Biography of *10:16.*
A Waverly trilobite **2**:69.
(and C. J. Herrick) Biological notes upon Fiber, Geomys and Erethizon **6**:15.
(and W. G. Tight) Central Nervous System of Rodents **5**:35.
Cerebrum and Olfactories of Opossum **6**:75.
Compend of Lithological Manipulation **1**:121.
Founded Bull. Sci. Labr. **10**:41.
Founded Den. Sci. Assoc. **10**:37.
Founded Journ. Comp. Neurol. **10**:54.
Geological History Lick. Co., O. **1**:4, **3**:13, **4**:11, 97.
(et. al.) Geology and Lithology of Michipicoten Bay **2**:119.
Limnicole or Mud-Living Crustaceae **1**:37.
Metamorphosis of Phyllopod Crustaceae **1**:15.
Notes on Carboniferous Trilobites **1**:51.

Osteology of Evening Grosbeak, Hesperiphona vespertina Bonap. **1**:5.
Rotifers of America **1**:43.
Hesperiphona, anatomy of **1**:15.
abeilii **1**:5.
vespertina **1**:5.
" specimen of **5**:22.
Heteranthera graminea **7**:87.
Heuchera americana **7**:36.
Hibiscus moscheutos **7**:24.
trionum **7**:24.
Hicks, Prof. L. E. Biography **10**:10.
principal writings **10**:10.
Hickory, Shell bark **7**:77.
Hieracium gronovii **7**:54.
paniculatum **7**:54.
scabrum **7**:54.
venosum **7**:54.
Hocking River and its tributaries **8**–II:52.
reversed drainage in **9**–II:37.
Hogweed **7**:50.
Honey-locust **7**:31.
Hop **7**:76.
Hop Clover **7**:29.
Hop Horn-beam **7**:78.
Hop-tree **7**:25.
Horehound **7**:70.
Horizontal component of the earth's magnetic force **2**:111.
Hornblende **1**:T15.
Horse Gentian **7**:44.
Horseradish **7**:18.
Horsetail **7**:99.
Horse Weed **7**:49.
Hound's Tongue **7**:61.
Houstonia caerulea **5**:10, **7**:45.
purpurea var. ciliolata **7**:45.
" " longifolia **7**:45.
Huckleberry **7**:56.
Humulus lupulus **7**:76.
Hyacinth **7**:84.
Hyalotheca mucosa **4**:132.
Hydra. Note on peculiar habit of fresh water, **4**:131.
Hydrangea arborescens **7**:37.
Hydrastis canadensis **7**:14.
Hydrocharidaceae **7**:80.
Hydrophyllaceae **7**:61.
Hydrophyllum appendiculatum **7**:61.
canadense **7**:61.
macrophyllum **7**:61.
virginicum **7**:61.
Hypericaceae **7**:22.
Hypericum ascyron **7**:22.

M

N

O

P

Q

R

Rush Creek, description of 9–II:
33–35.

Rutaceae 7:25.
Rutile 1:T4.

S

Sabbatia angularis 7:59.
Sagenite 1:T4.
Sagittaria variabilis 7:89.
   variabilis var. angustifolia
           7:89.
   "   " gracilis 7:89.
   "   " · latifolia 7:89.
   "   " obtusa 7:89.
   heterophylla, var. angustifolia
           7:89.
   "   " rigida 7:89.
St. John's–wort 7:23.
Salicaceae 7:79.
Salite 1:T14.
Salix alba, var. vitellina 7:79.
   cordata 7:79.
   discolor 5:10, 7:79.
   fragilis 7:79.
   humilis 7:79.
   longifolia 7:79.
   nigra 7:79.
   purpurea 7:79.
   sp. 5:10.
Salpina 1:52.
   affinis n. sp. *1:52.*
   brevispina 6:60.
   mucronata 1:52, 6:60.
Salt Creek, Preglacial drainage in
          9–II:28.
Sambucus canadensis 1:30, 7:43.
Samolus valerandi, var. americanus
          7:57.
Sandstone 9–I:20.
   Waverly 9–I:21.
Sandwort. Thyme–leaved 7:21.
Sanguinaria canadensis 7:16.
Sanguinolites aeolus *3:70.*
   amygdalinus 3:69.
   contractus *3:69.*
   flavius *3:69.*
   marshallensis *3:67.*
   michiganensis *3:70.*
   naiadiformis *3:71.*
   nobilis, *see* Allorisma nobilis
    3:71.
   obliquus 3:70.
   rigida, *see* S. transversus
          3:68.
   senilis n. sp. 3:66.
   transversus 3:68.
   unioniformis *3:67.*
Sanicula marylandica 7:42.
      " var. canadensis
          7:42.
Sanidine 1:T13.

Santalaceae 7:75.
Sapindaceae 7:27.
Saponaria officinalis 7:21.
Sarsaparilla 7:42.
Sassafras officinalis 7:74.
Sault Ste. Marie 1:5.
Saxifragaceae 7:36.
Saxifraga pennsylvanica 7:36.
   virginiensis 7:36.
Saxifrage, Swamp 7:36.
Scale divider 5:20.
Scapholeberis mucronata 6:67, 8–I:3.
Scapolite 1:T5.
Scaridium longicaudum 6:60.
Scarlet Oak 7:78.
Scenedesmus polymorphus 2:115.
   quadricauda 2:115.
Scheuchzeria palustris 8–II:5.
Schist, Chloritic 2:126.
   near Pacolet Mills S. C. 4:8.
Schizodus affinis sp. (?) n. *2:41.*
   amplus 2:41, 43.
   chemungensis, var. aequalis
          3:64.
   cuneatus 2:41, *42.*
   cuneus *3:65.*
   curtis 2:41, *42,* 2:145.
   harlamensis n. sp. *4:117.*
   medinaensis *3:65.*
   newarkensis n. sp. *3:64, 4:36.*
   occidentalis 2:41.
   palaeoneiliformis n. sp. *3:96,*
    4–Pl. 6.
   perelegans 2:41.
   prolongatus n. sp. *4:36.*
   quadrangularis 4–Pl. 6.
   (?) spellmani n. sp. *2:42.*
   subcircularis n. sp. *2:41,*
          2:145.
   triangularis *4:116.*
   wheeleri *2:42.*
Scientific publications 10:41.
   terms, Pronunciation of,
    Latin and Greek 4:161.
Scioto Basin, abnormal drainage in
    9–II:20.
   River 9–I:15.
    character of its upper
     and lower portion
         8–II:53.
   Valley, relation to Ohio Val-
    ley 9–II:25.
Sciotoville, O., geological section
       near 4:102.
Scirpus atrovirens 7:91.

U

V

W